International Conference on Advanced Materials Behaviour & Characterization (ICAMBC 2021)

IOP Conference Series: Materials Science and Engineering
Volume 1170

Online
24-26 April 2021

Editors:

I. Saravanan
S. Arulvel

ISBN: 978-1-7138-4630-7
ISSN: 1757-8981

Printed from e-media with permission by:

Curran Associates, Inc.
57 Morehouse Lane
Red Hook, NY 12571

Some format issues inherent in the e-media version may also appear in this print version.

Additional copies of this publication are available from:

Curran Associates, Inc.
57 Morehouse Lane
Red Hook, NY 12571 USA
Phone: 845-758-0400
Fax: 845-758-2633
Email: curran@proceedings.com
Web: www.proceedings.com

TABLE OF CONTENTS

IOP Conf. Series: Materials Science and Engineering 1170 (2021) 011001 doi:10.1088/1757-899X/1170/1/011001

PREFACE

The volume of the content has been published through ICAMBC_2021 (International Conference on Advanced Materials Behavior and Characterization_2021) conducted as virtual online conference on 24-26, April 2021. The conference was organized by Mattest Research Academy, Chennai, Tamil Nadu, and India.

The focus of the conference is to promote the innovative works from core and diversified fields of materials engineering for different applications. Materials engineering is one of the very most important interdisciplinary fields that focuses on studies fundamental physical and chemical basis in order to develop new materials and compounds through manipulating atoms to form desired structures and properties. Any revolutionary changes are always through the breakthrough in new materials.

The manuscripts were submitted to the conference from multidiscipline of Materials, Engineering and Sciences. The authors and participants from international universities and National institutions and Research organizations were joined through online via Google Meet. The experts and session chairs have provided talk related to recent research innovations and importance of the virtual conference in this pandemic situation due to COVID'19.

The forum comprised around six invited talks in the area of Bio sciences Medical sciences Recent Advanced Materials such as Nano Materials Polymers, Ceramics Natural Fiber Reinforced Composites, Phase Change Materials, Surface Engineering, tribology and biomaterials.

The conference had received a good opening to the research society and everyone encouraged about the event was conducted for three days in pandemic situation. Each presentation is around 20-30 minutes as parallel sessions in front of subject experts. The interaction session provided the real experience to the participants and they had good exposure with the students and delegates from global Institutions. The global participants from all states in India, South Africa, Nigeria, Iran, Turkey, Morocco Ethiopia, Bangladesh, and Canada had presented and participated in the conference. Indeed it was a wonderful experience for the global researchers to share their research innovations and the feedback is the evidence from the participants registered in our website. https://www.mattest.net/feed-back

Dr.I.Saravanan
Convener-ICAMBC_2021
Mattest Research Academy, Chennai, Tamil Nadu, India.

IOP Conf. Series: Materials Science and Engineering 1170 (2021) 011001 doi:10.1088/1757-899X/1170/1/011001

Editors

1	Dr.I.Saravanan	Professor, Mattest Research Academy, Chennai, Tamil Nadu, India.
2.	Dr.S.Arulvel	Assistant Professor (Sr), VIT University, Vellore, Tamil nadu, India.

Committee Members

Advisory Committee		
1	Prof. Mannar.S.L,	AICTE-Distinguished Visiting Faculty, Director (Former) , MMG, IGCAR
2	Prof. M. Kamaraj,	Dept. of Metallurgical and Materials Engineering, IITM, Chennai, India.
3	Prof. V. Balasubramanian	Head and director in CEMAJOR, Dept. of Manufacturing Engineering, Annamalai University, Annamalai Nagar, India.
4	Prof. A. Elayaperumal	Head, Engineering Design Devision,Anna University, Chennai, India.
5	Prof. B.Suresha,	Head, Dept. of R&D NIE,Mysore, India.
6	Prof.Wang Liming.	Associate professor, School of Mechanical Engineering, Shandong University, Jinan, China
Technical Reviewer Committee		
1	Prof. P.Ramkumar,	Associate Professor, IITM, Chennai, India.
2	Prof. N.Venkateshwaran	Head, Dept. of Mechanical Engineering, Rajalakshmi Engineering College, Chennai, India.
3	Dr.P.Vijayakumar,	Associate professor, Anna University, Kancheepuram. Tamil Nadu, India.
4	Capt. Dr. M.Seeman	Associate Professor, Dept. of Manufacturing Technology, Annamalai University, Annamalai Nagar, Tamil Nadu, India.
5	Dr. Hammadi EL FARISSI	Department of Chemistry,Laboratory of Chemical Engineering and Valorization of Resources, Faculty of Sciences and Technologies, Abdelmalek Essaadi University, Tangier, Morocco.
6	Dr.FOUADEL MANSOURI	Department of Chemistry,Laboratory of Chemical Engineering and Valorization of Resources, Faculty of Sciences and

		Technologies, Abdelmalek Essaadi University, Tangier, Morocco.
7	Dr.Venkatesa Prabhu S	College of Biological and Chemical Engineering, Addis Ababa Science and Technology University, Addis Ababa, Ethiopia.
8	Dr.S.Sathish	Head Dept. of Mechanical Engineering, Aalim Muhammed salegh college of engineering, Chennai, Tamil Nadu, India.
9	Dr..J.Sandeep	Assistant Professor, Department of Printing Technology, Anna University, Chennai, Tamil Nadu, India
10	Dr. U.Mohamed Iqbal	Associate Professor, Department of Mechanical Engineering, SRM Institute of Science and Technology, Chennai, Tamil Nadu, India
11	Dr. S. Palani	Professor, Department of Mechanical Engineering, Vel Tech Multi Tech Dr. Rangarajan Dr. Sakunthala Engineering College, Avadi, Chennai, Tamil Nadu, India
12	Dr. B.Ramesh	Professor, Institute of Mechanical Engineering, Saveetha School of Engineering, Saveetha Institute of Medical and Technical Sciences, Chennai, Tamil Nadu, India.
13	Dr. N. Yuvaraj	Associate Professor , Department of Mechanical Engineering - Vel Tech Multitech Dr. Rangarajan Dr. Sakunthala Engineering Avadi Chennai, Tamil Nadu, India.
14	Dr.Gopi Saravanan	Scientist, Assistant Professor, Centre for Nanoscience and Nanotechnology, International Research Centre, Sathyabama Institute of Science and Technology,Chennai, Tamil Nadu, India.
15	Dr. Vaishali Deshmukh	Assistant Professor, Department of Physics, Shri Shivaji Science College, Shivaji Nagar, Amravati, India.
16	Dr. M. Nithya	Associate Professor, Department of Civil Engineering, CMR Institute of Technology, Hyderabad, India.
17	Dr.M.Puviyarasan	Associate professor, Dept. of Mechanical Engineering, Panimalar engineering college, Chennai, Tamil Nadu, India
18	Dr.S.Madhu	Associate Professor, Dept.of Automobile Engineering, Saveetha school of Engineering, Chennai, Tamil Nadu, India
19	Dr. N. Nithyavathy	Associate Professor, Department of Mechatronics Engineering, Kongu Engineering College, Perundurai, Erode, TamilNadu, India.
20	Mrs. I. Srilega	Asst. Professor, M.Sc (Chemistry), M.Sc (Bio Chemistry), M.Phil, Dept. of Bio Chemistry, SIVET College, Chennai, Tamil Nadu, India
21	Mrs. P. Samatha chowdary	Dept. of Civil Engineering, R. V. R & J. C College of engineering, Guntur, Andhra Pradesh.
22	Dr. Vinothkumar Sivalingam	Associate Professor, School of Mechanical Engineering, Shandong University, Jinan, China.

23	Dr. M.S.Jagadeeshwaran	Associate professor, SA Engineering College, Chennai, Tamil Nadu, India
24	Mr.Gopi	Assistant Professor, Department of Mechanical Engineering, Adhi College of Engineering& Technology, Chennai, Tamil Nadu, India.
25	Mr.P.Kumaravel	Assistant Professor, Department of Mechanical Engineering, Adhi College of Engineering& Technology, Chennai, Tamil Nadu, India.
26	Mr. Sai Krishnan	Department of Mechanical Engineeering, Rajalakshmi college of Engineering, Chennai, Tamil Nadu, India.
Organizing Committee		
1	Mr. S. Manoj Kumar	Assistant Professor, Dept. of Mechanical Engineering, JNN Institute of engineering and Technology, Chennai.
2	Capt. P. Ravichandran	Dept. of Civil Engineering, Annamalai University, Annamalai Nagar.
3	Mrs.E.Nirmala Devi	Assistant Professor of Physics, G. Pulla Reddy Engineering College, Kurnool, Andhra Pradesh, India.
4	Ms.Karishma Maheshwari	Department of Chemical Engineering, Malaviya National Institute of Technology, Jaipur.
5	Mr. M. Karthikeyan	Technical Consultant, Mattest Research Academy, Chennai, Tamil Nadu, India.

Peer review declaration

All papers published in this volume of IOP Conference Series: Materials Science and Engineering have been peer reviewed through processes administered by the Editors. Reviews were conducted by expert referees to the professional and scientific standards expected of a proceedings journal published by IOP Publishing.

- **Type of peer review:** Double Blind Review
- **Conference submission management system:** Submission has been received through mail and Peer Review has been followed.
- **Number of submissions received:** 27
- **Number of submissions sent for review**: 27
- **Number of submissions accepted:** 14
- **Acceptance Rate**

  (Number of Submissions Accepted / Number of Submissions Received X 100): 51.85%
- **Average number of reviews per paper: 3 (1-Preliminary-** Format, Language, Reference Cross Check and Title search **+ 2-Technical)**
- **Total number of reviewers involved:** 30

Any additional info on review process:

- All the manuscripts have been reviewed by subject Experts from the various potential research Institutions. Focused on language, Format, Similarity and Novelty. And the revised manuscript will be incorporated as per the suggestion and comments from the journal.

- **Contact person for queries:**

Dr.I. Saravanan,
Convener & Guest Editor,ICAMBC_2021.
Email : is.mech@outlook.com
Phone : 98414 44418

A surface modified microbial polymer inactivates quorum sensing molecules and incapacitates Sphingomonas paucimobilis biofilm formation in plumbing material

Parul Gulati[1] and Moushumi Ghosh [2]

[1] Department of Biotechnology, Thapar Institute of Engineering and Technology, Patiala, India
[2] Department of Biotechnology and affiliate faculty, Centre of Excellence Emerging Materials, Thapar Institute of Engineering and Technology, Patiala,India

E-mail: mghosh@thapar.edu

Abstract: We report the potential antibiofilm properties of a chemically modified biopolymer on the formation of biofilms by Sphingomonas paucimobilis MG6 an opportunist pathogen isolated from distributed drinking water systems. The biopolymer obtained from the producer bacteria K. terrigena was quarternized and examined for safety using RAW 264.7 cell lines. The modified biopolymer (N-methyl biopolymer: NMB) at 1mg/ml inactivated AHL homologs of S. paucimobilis at ambient temperature.GC-MS and C. violaceum biosensor assays were used to evaluate loss of AHL molecules. On NMB coated coupons loss of biofilm formation under static conditions, was visualized by a rapid calcofluor assays and scanning electron micrographs (SEM). Applicability was further investigated in a flow through apparatus afforded with different plumbing materials, coated with NMB.Leaching of NMB was not observed and a complete absence of biofilms was noted on coated materials, especially in PVC. SEM and viable counts corroborated these observations. Overall, results of this study suggest that the novel material offers as a potential sustainable approach for controlling biofilms in drinking water systems.

Keywords: Antibiofilm, Biopolymer, N-methylbiopolymer, plumbing, quorum sensing, S. paucimobilis

1. Introduction

Sphingomonas, considered as opportunistic pathogen [17],[25],[28] is endowed with the capability to survive in oligotrophic niches important for human life and sustenance-such as water supply, reverse osmosis systems[32]. Biofilm formation by Sphingomonas on a diverse range of surfaces [10], [11], [31] including plumbing materials and water distribution systems has been demonstrated [10]. Biofilm formation on these surfaces has been linked to bacteremia, catheter infections (intravascular), infections in the urinary and biliary tract, skin, ventilator associated pneumonia, myositis [27], diarrheal disease [7] in dental unit waterlines[26], tap water [12] and in hospital water system [27]. Sphingomonas paucimobilis has been frequently implicated in the above and in addition as a candidate for evoking metal corrosion and biofouling of distribution systems used for water [19].

Chlorine and biocide resistance of this strain exacerbates the risk to consumers.

To prevent the detrimental effects of biofilm produced by Sphingomonas, it is indispensable to employ novel strategies for inhibition of biofilm. Since bio film formation is governed by quorum sensing, interference and blocking quorum sensing signals responsible for biofilm formation can be an effective, non-disruptive strategy. Biopolymers with several novel functions have gained importance as surface applications in recent times. Their role in interference with quorum sensing signals (N-acyl-homo-serine lactones (AHLs)) in gram negative bacteria can be important in circumventing biofilms in water distribution systems without evoking residual after effects. Polymers originating from biological sources are used in a wide range of applications due to unique structural and functional attributes [20]. Moreover biopolymers have intrinsic safety, amenability of structural modification and have been used for water purification. The present study attempted to elucidate the role of a biopolymer from K. terrigena with quorum signal binding activity responsible for biofilm formation by S. paucimomilis. To simulate the applicability in distributed water systems different plumbing materials were coated with the surface modified biopolymer and analyzed for biofilm production in a flow through system previously reported.

2. Materials and methods

2.1. *Microorganism and culture conditions:*

Overnight grown cultures of Sphingomonas paucimobilis MG 6, were used to produce biofilms on plumbing materials as described [9]. Extracellular biopolymer was obtained from K. terrigena (Accession number EU082029) and purified and quarternized [14]. Briefly, the biopolymer (500 mg; 10 mmole) was mixed with 1% (v/v) acetic acid (50 ml) and formaldehyde (0.5 ml; 0.22 mmole) was added. After stirring, the suspension at room temperature for 30 min, sodium borohydride (41.25 mg; 1.12 mmole) was added. The pH of the suspension was then ad j u s t ed to 10 by NaOH (1 M). White precipitates were filtered and washed until they attained a pH of 7. Azeotropic distillation was used to remove trapped moisture was removed by refluxing with dry benzene. Evaporation of benzene resulted in N-methylbiopolymer (NMB) (yield 86%; 430 mg). The cytotoxic activity and cell viability of the modified biopolymer was analyzed using RAW 264.7 macrophages and MTT assay [13]. RAW 264.7 cells (National Centre for Cell Sciences, Pune, India) were propagated in Dulbecco's modified Eagle's medium (DMEM) with 10% fetal bovine serum (FBS), 1% penicillin-streptomycin solution, 1% L-glutamine and HEPES[4-(2- hydroxyethyl)-1-

piperazineethanesulfonicacid]. A 5% carbon dioxide concentration and incubation temperature of 37°C was used during the experiment. Plates with 96 wells were afforded with cells (50,000 cells/0.2 mL/well), incubated overnight for attachment; the medium in wells was replaced with NMB-containing medium the following day. A stock solution of 1 mg/mL was prepared in sterile saline, suitably diluted in culture medium to obtain working solutions (0.2, 0.4, 0.6, 0.8 and 1 mg/L) and 0.2 mL was added. For cell viability measurements, cells were incubated for 24 or 48 hours and the MTT assay as described earlier [33] was used.

2.2. *Biofilm Inhibition assay:*

Twenty microliters of inoculum (optical density, O.D.600 =0.02) was added along with 0.03 mg/ml NMB in each well of the polystyrene plates and incubated at 37ºC for up to 72 hours. The plates were removed every 4 hours and biofilm formed was measured by crystal violet assay. The percentage of biofilm inhibition was calculated by:

$$Percentage\ of\ Biofilm\ inhibition = \frac{[Control\ O.D\ 585\ nm - Test\ O.D\ 585\ nm]}{Control\ O.D\ 585\ nm} * 100$$

2.3. *Simulation of biofilm inactivation in flow through system*

A laboratory-scale simulated distribution system [9] was fabricated with minor modifications. Different sets of pipe material with the exception of Copper and were used. These included: polyvinyl chloride (PVC), Polypropylene (PP), polyethylene (PE), Aluminum (Al), and rubber (R). The distribution pipe has a length of 1 m and an internal diameter of 3 cm. Tap water was continuously pumped from the contact tank (30 liters of tap water) at a flow rate of 1.0 L/min into the simulated drinking water distribution pipe using a peristaltic pump. Overnight grown S. paucimobilis MG6 culture was used as inoculum, the concentration of viable cells was 10^5CFU/ml of water pumped in the contact tank.

2.4. *Biofilm inactivation on coupons with and without modified biopolymer*

To check the inhibition of biofilm formed on plumbing materials under static conditions, NMB was added to sterile coupons and dried for 1 hour. S. paucimobilis MG6 was inoculated on all the coupons in water and Sphingomonas selective growth (SSG) medium and biofilm was allowed to form for 72 hours. These were then removed using sterile forceps, washed with 50 ml of 1xPBS to free unadhered cells. 80 mg/l calcofluor was added for staining and washed with PBS to remove unbound dye. The presence of biofilm on the surfaces was visualized by a hand-held UV-lamp long wave (Blak-Ray Lamp model UVL- 2 1, UV-366 nm)[19].

To check for leaching of NMB, coupons as well as plumbing materials were pre coated with different concentrations of NMB ranging from 0.1 to 1 mg/ml by dip coating method. The coupons were allowed to dry for 1 hour and were placed in water and media. The coupons and materials were continuously analyzed for 72 hours for leaching of NMB using calcofluor assay.

2.5. Scanning Electron Microscopy (SEM) of PVC coated with modified biopolymer

Biofilm of S. paucimobilis MG6 developed on PVC coated with and without NMB was examined after 3 days by scanning electron microscopy (SEM) [1]. The samples of biofilms were fixed using glutaraldehyde solution (2.5%) overnight at 4°C. Excess glutaraldehyde was removed by washing was carried with 1x Phosphate Buffer Saline (PBS). Dehydration of biofilm samples was performed with 30% ethanol followed by 50%, 60%, 70%, 80%, 90%, 95% to 100% ethanol respectively. Gold coating of samples were done prior to analysis by SEM (JSM541-V, JEOL, Japan) [15].

2.6. AHL Extraction

The OD600 of overnight grown S. paucimobilis MG6 was adjusted to O.D 600~ 1.80. The culture was centrifuged at 10,000 rpm for 15 minutes and cell pellet was discarded. The supernatant was filtered through 0.2 µm membranes. The filtrate was mixed with acidified ethyl acetate (0.01% Glacial Acetic Acid) and extracted for 15 minutes. Mixture was set aside for 10 minutes in a separating funnel for obtaining upper organic and lower aqueous layer. After extracting the aqueous layer twice the entire *organic* layer was pooled and dried (Rota-vapour) under vacuum at 40-50ºC. The dried residues were dissolved in1 ml of 20% acetonitrile.

2.7. Well diffusion assay for quorum sensing inhibitory activity:

A simple well diffusion assay was applied for detecting the presence of N-Acyl Homoserine Lactone (AHL) following incubation with NMB, in S. paucimobilis MG6. The biosensor strain have been used extensively; C. violaceum 026 is the mini Tn5 double mutant defective in the synthesis of violacein pigment, but retains the ability to respond to C4-C6 AHLs. Five milliliter of molten semi-solid Luria agar (0.3% w/v) was inoculated with 50 µl of an overnight grown culture of C. violaceum 026 and immediately poured over the surface of LB agar plates. When the overlaid agar solidified, wells were punched in the agar and filled with the cell-free supernatant and adjusted to a final solution of 50 µl by addition of sterile LB. Petri dishes were incubated overnight (28°C), then analyzed the same for induction or inhibition of synthesis of violacein [23]. Purified AHL incubated with the NMB was analyzed in parallel for biosensor activity.

2.8. Gas Chromatography-Mass Spectrometry

Purified S. paucimobilis MG6 AHL (incubated) with and without NMB was examined by gas chromatography-mass spectrometry (GC-MS) (PerkinElmer Clarus 680 GC, India) [5]. Samples were injected into ELITE-5MS capillary column (30.0 m, 0.25 mm, 250 μm df). The operation parameters used were: Helium as carrier gas; injector temperature: 260°C; 1 ml/min flow rate. The temperature of the oven temperature was programmed as: 60 °C (2 min); followed by 300 °C at the rate of 10 °C /min and 300 °C, where it was held for 6 min. MS detector conditions were: transfer line temperature 240 °C; ion source temperature >240 °C; and ionization mode impact at 70 eV.

3. Statistical analysis: Data was converted to mean ± standard deviation [15] analyzed by one- way analysis of variance (ANOVA). In addition this was followed by Tukey's test or two-way ANOVA wherever applicable. The Graph Pad Prism was applied and results were considered significant for $p<0.05$.

4. Results

The survival of RAW 264.5 macrophages was not affected at 24 hours of incubation in the presence of NMB (modified biopolymer) significantly (p>0.05) (Table 1).

Table 1 Impact of NMB (0.02-1mg/L) on viability of RAW 264.7 cells following 24 hours incubation. Data shows mean±SD of two experiments (performed in triplicate). Significant changes (p>0.05)(not shown) were not observed after 48 hours

Biopolymer dose (mg/L)	Cell Viability (%)
0.00	100% ±3.27
0.02	99.78% ± 4.11
0.40	98.97% ± 7.10
0.60	99.03 % ± 5.32
0.80	99.64 % ± 7.31
1.00	98.67% ± 6.76

The modified biopolymer treated macrophages upon microscopic observation, was visible as a monolayer culture without any distinctive morphological changes at bactericidal doses over an incubation period of 48- 72 hours. The modified biopolymer (NMB) was adjudged safe from these results.

Biofilm formation of S. paucimobilis MG6 (Accession No KX594380) was studied by crystal violet assay. We observed sloughing of biofilms usually by 96 hours therefore the 72 hours study was considered more relevant. Sloughing of biofilms has been reported in plumbing materials recently [7] in line with our observations.

Figure 1 depicts the influence of NMB on biofilm progression over time.

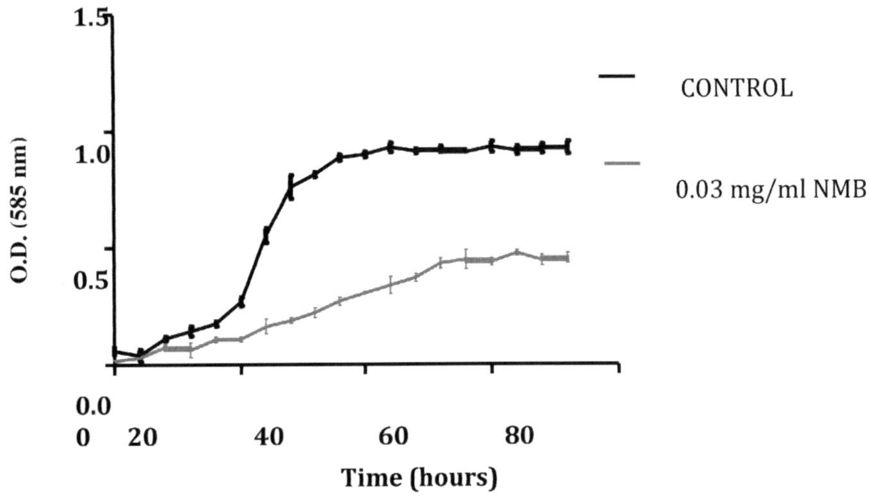

Figure 1: Inhibition of biofilm formation by S.paucimobilis MG 6 observed with NMB.Complete loss of biofilm was observed at NMB concentration of 1mg/ml. Data is the mean±SD of two experiments performed in triplicate.

Leaching of NMB from the coated coupons or plumbing materials used was not observed during the tenure of the study (96 hours). Biofilm production was reduced sharply and with lower (0.03mg/ml) doses of NMB and 45% biofilm reduction was noted. At a concentration of 1mg/ml, complete loss of biofilm was observed; therefore this concentration was used for coating the plumbing materials. Although evaluation of the physical effects during biofilm formation was not our objective, it was important to understand whether the different plumbing materials coated with NMB, actually supported biofilm formation by S. paucimobilis MG6 in the same fashion as observed under static conditions. The observations of the flow through simulation experiment corroborated the fact that NMB coated plumbing materials failed to support biofilm development, albeit with different efficacies. There was a complete loss of biofilm formation on PVC; in PP, PE and Al notable decrease in biofilm formation was observed. Viable cells of S. paucimobilis MG6 were observed from different scrapings of plumbing materials obtained at various ages (Table 2).

Table 2: Comparison of biofilm cells log counts which developed on different plumbing materials with and without NMB treatment. PVC: polyvinyl chloride; PP: polypropylene; PE: polyethylene; Al: Aluminum. *The values represented are log CFU /ml of S. paucimobilis MG6

Tested plumbing materials:	**Biofilm age** **24 hours**	⇨ **48 hours**	**72 hours**	**96 hours**
PVC **Untreated** **NMB treated**	**90.2** **Not detected**	**93.6** **Not detected**	**94** **Not detected**	**95.3** **Not detected**
PP **Untreated** **NMB treated**	**87.4** **Not detected**	**88.0** **Not detected**	**89.0** **Not detected**	**90.4** **3.6**
PE **Untreated** **NMB treated**	**82.5** **Not detected**	**83.3** **Not detected**	**86.7** **4.3**	**79.4** **3.3**

Al				
Untreated	**74.7**	**76.0**	**81.0**	**81.4**
NMB treated	**Not detected**	**Not detected**	**Not detected**	**2.5**

We observed bacterial growth at 72 and 96 hours on PP, PE and Al. The surface composition of the surface is pertinent in sustaining biofilms [26]. Therefore the surface characteristics of these plumbing materials, following 48 hours of exposure to S. paucimobilis (in the flow through system) should provide an explanation. PVC is extensively used as a plumbing material for distribution of water, loss of biofilm production in NMB coated PVC thus assumes relevance. Since intensity of fluorescence reciprocates biofilm formation, the effect of NMB on biofilm inhibition could be clearly visualized (Figure 2).

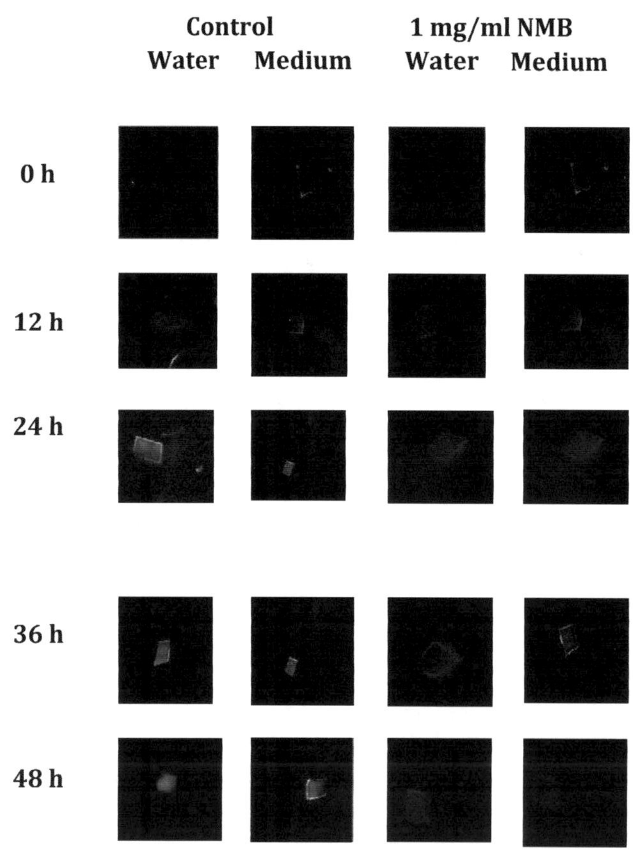

IOP Conf. Series: Materials Science and Engineering 1170 (2021) 012001 doi:10.1088/1757-899X/1170/1/012001

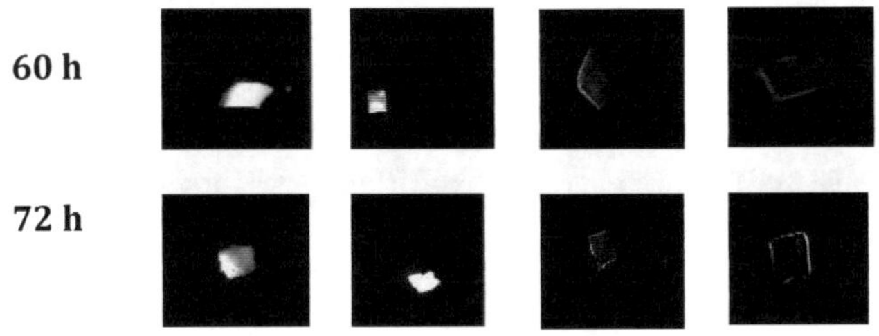

Figure 2: Inhibition of S. paucimobilis MG6 biofilm on PVC coupons in water and medium. Presence of biofilms was analyzed using the calcofluor assay.

Distinct reduction in biofilm formation in the NMB coated coupons was also observed from scanning electron micrographs confirming the biofilm inhibition efficacy (Figure 3).

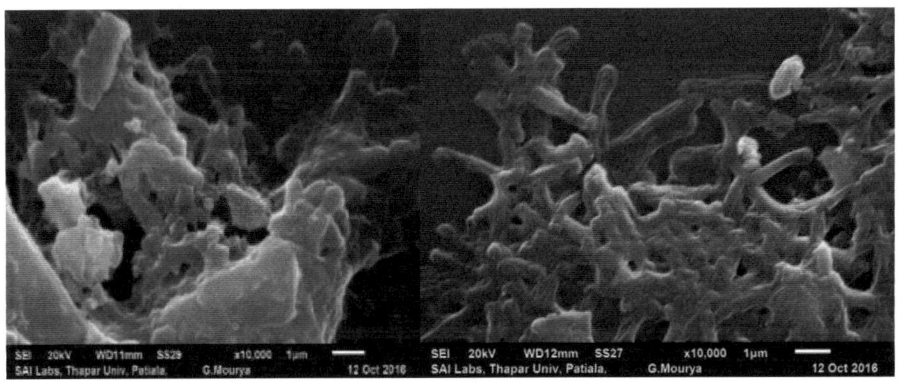

CONTROL **NMB treated**

Figure3: Scanning electron Micrographs of S. paucimobilis MG6 (control, Left) with 1 mg/ml NMB coated on PVC and used in the flow through system.

It has been suggested that for biofilm formation, AHL signaling is important. However, there is little evidence for AHL regulation of biofilm development in Sphingomonas. Therefore an attempt was made to detect AHL in the isolate. For this, the CV026 inhibition test was carried out since these molecules are expected to inhibit the induced CV026 and produce colorless colonies in purple background. Inhibition of AHL activity was observed from the diminished violacein production in C. violaceum CV026 in a concentration dependent manner.

Moreover, this particular assay is invaluable for ascertaining the acyl chains ranging from C10 to C14 [22]. The A. tumefaciens A136, a broad range biosensor, was also employed to confirm the presence of the C8 to C12-HSLs including oxo-C6-HSL; the good sensitivity to even low level of longer acyl side chain AHLs [8] is notable, for this biosensor assay. To demonstrate whether NMB can degrade AHL signaling molecule of S. paucimobilis MG6, AHL incubated with NMB was used for evaluating its functionality by the biosensor. Chemical analysis for the presence of AHL upon treatment with NMB was carried out by GCMS. After incubation of NMB with dDHL, it was observed that [M-H] ions which are characteristic of this signaling molecule are absent, while the presence of specific ions at m/z 69, 86 and 128 confirmed the loss of dDHLs(Figure 4).

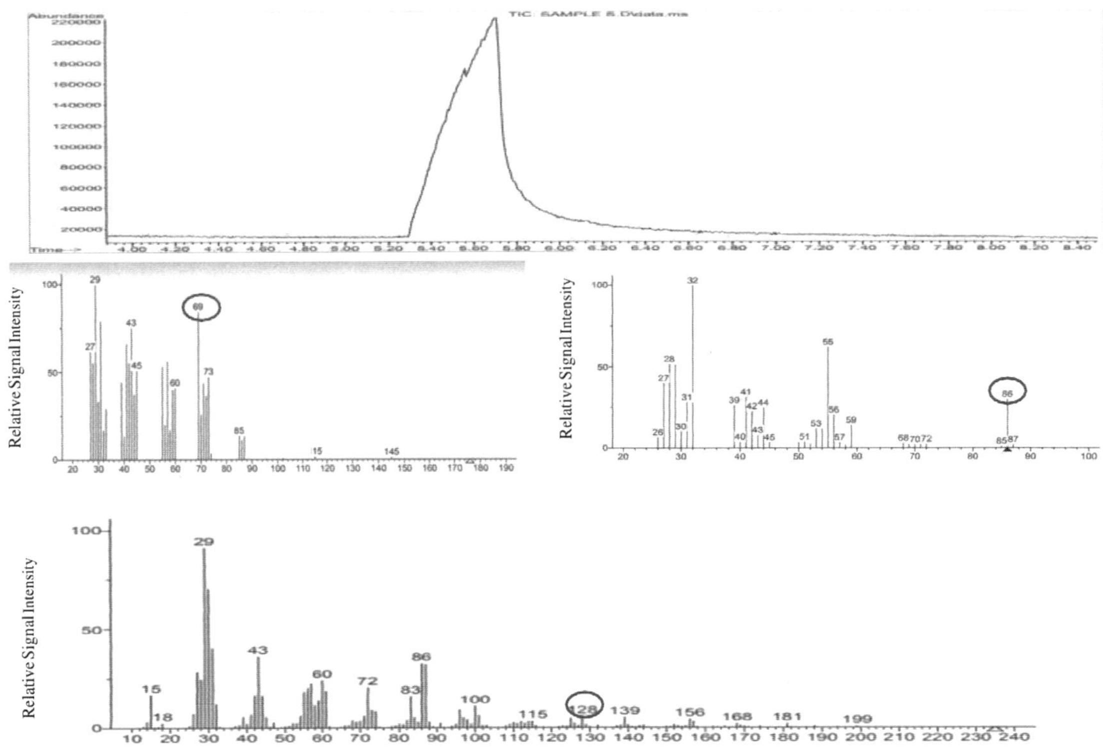

Figure 4: Spectra of extracted AHL (N-dodecanoyl-L-Homoserine lactone) degradation by NMB in full scanning positive mode (a) Total Ion Chromatogram (b)-(d) Ion-fragments at different m/z carried out by GC-MS.

These results indicated that NMB possibly binds to AHL molecules and subsequently degrades them.

5. Discussion

Earlier studies in this laboratory have explicitly demonstrated the capability of S. paucimobilis MG 6, an isolate from community drinking water systems, to form biofilms on a wide variety of plumbing materials. The strain is chlorine tolerant and supports multigeneric aggregation with S. flexneri 2a and E. coli 0157:H7.Furthermore a systematic study on mechanisms of biofilm formation using biofilm negative mutants conclusively highlighted the involvement of quorum signaling molecules in Sphingomonas for biofilm formation. A complete loss of motility, aggregation along with exopolysaccharide production was noted in biofilm negative mutants. The latter properties are revoked upon supplementation with homolog of N-dodecanoyl-L- Homoserine lactone purified and characterized from S. paucimobilis MG6. Incubation with purified AHL with the modified biopolymer led to a complete loss of swarming motility and quorum sensing activity detectable on A. tumefaciences plate assays. It been suggested that Interference of the bacterial communication system by suitable antagonists can offer a benign yet effective strategy for biofilm prevention. To this end, microbial exopolymers are attractable classes of molecules and present an interesting proposition for sustainable application. The involvement of AHLs in biofilm formation by Sphingomonas assumes importance, in view of the safety of drinking water. Thus, it was the intent of this study to investigate the role of bacterial exopolymers modified chemically. The biopolymer from Klebsiella terrigena *has* been well characterized in our prior studies, besides the native biopolymer is able to inactivate to purified AHL but with very low efficiency. This biopolymer was therefore selected for chemical modification and designated as N-methyl biopolymer (NMB). Parallel results (not shown) suggest that NMB modulated the quorum sensing genes luxI and *luxR* activities leading to a loss of biofilm formation without affecting the cell integrity presumably by lactonolysis of lactone moiety of signal molecule. This results in the ring opened product and complete inactivation of the signal molecule crucial for biofilm formation.

Thus far, several studies have reported synthetic, semi semi-synthetic and natural molecules with antiquorum sensing activities. However, to date, no studies have evaluated antiquorum sensing molecules especially in materials used for plumbing or water distribution. The conditions of the environment define the growth of biofilms in biotechnological systems, in the human body, and in nature. Growth of microbes is strongly mediated by the nutrient availability, physical transport, and structure of biofilm [4], [16], [35]. For instance, it has been observed that biofilms develop different morphologies under different flow conditions [3], [6], [29], [11], [34] and are strongly affected by hydrodynamic shear forces in wastewater treatment plants [21], and internal transport limitations are among many factors that cause biofilms to be highly resistant to chemical stresses[16],[18]. It is clear that the flow distributions, hydrodynamic shear on surfaces in addition to important chemical constituents govern habitat conditions for microbial growth in biofilm [2], [15], [18], [30]. Several flow through systems has been experimented in an effort to unravel interactions between the biofilm growth and their spatial patterns evoke by key environmental conditions [8]. In the present study it was important, first, to understand that whether the modified biopolymer could actually support biofilm formation by S. paucimobilis MG6 under static conditions. These observations were then correlated with the flow through simulation experiment. Viable cells of S. paucimobilis MG6 were recorded from different scrapings of plumbing materials at different ages (Table 2), direct staining using calcoflour and SEM images clearly indicated the functionality of the modified biopolymer in restricting biofilm formation. Violacein production is a simple assay for the detection of AHLs. Since, violacein is inducible by all the AHL (with varying N-acyl side chains from C4 to C8 in length) In CV026 [24]. Therefore this strain was deemed suitable for use as bioreporter for directly observing the effect of AHL treated with and without NMB in the present study. The bio-reporter studies and analytical results of direct incubation of AHL with NMB clearly substantiate the observations on inactivation of AHL which is translated to an impaired biofilm formation by S. paucimobilis.

To sum up, our findings suggest that the modified biopolymer can effectively inhibit biofilm formation by S. paucimobilis especially on the surfaces indicated. The flow through system enabled a simulation of real time conditions. Although stricter validations will be necessary for enabling suitable modifications prior to commercial applications, the current results provide important insights for developing an effective and alternate strategy using this biopolymer to prevent biofilms in distributed water systems.

6. Conclusion

Overall, the present study demonstrated the applicability of NMB, a quarternized extracellular polymer produced by K. terrigena in inactivating AHL of S. paucimobilis. In absence of AHL molecules, S. paucimobilis fails to produce biofilm. The anti-biofilm effect could be replicated directly on plumbing materials under simulated conditions using a simple flow though setup. The results provide encouraging insights for further studies to exploit the applicability of NMB as a novel green material for controlling Sphingomonas biofilms in distributed water systems.

References:

[1] Asahi Y, Miura J, Tsuda T, Kuwabata S, Tsunashima K, Noiri Y, Sakata T, Ebisu S, Hayashi M 2015 Simple observation of Streptococcus mutans biofilm by scanning electron microscopy using ionic liquids *AMB Exp.***5** 6-8.

[2] Battin TJ, Wille A, Sattler B, Psenner R 2001 Phylogenetic and functional heterogeneity of sediment biofilms along environmental gradients in a glacial stream *Appl Env.Microbiol.***67**

799- 807.

[3] Besemer K, Singer G, Limberger R, Chlup A-K, Hochedlinger G, Hödl I, Baranyi C, Battin TJ 2007 Biophysical controls on community succession in stream biofilms. *Appl. Env. Microbiol* **73** 4966-4974.

[4] Boessmann M, Neu T, Horn H, Hempel D 2004 Growth, structure and oxygen penetration in particle supported autotrophic biofilms *Wat. Sc. Tech.***49** 371-377.

[5] Burton EO, Read HW, Pellitteri MC, Hickey WJ 2005 Identification of Acyl-Homoserine Lactone Signal Molecules Produced by Nitrosomonas europaea Strain Schmidt *Appl. Env. Microbiol.*7**1** 4906-4909.

[6] Chang YC, Le Puil M, Biggerstaff J, Randall AA, Schulte A, Taylor JS 2003 Direct estimation of biofilm density on different pipe material coupons using a specific DNA-probe. *Mol. Cellu.Prob.***17** 237-243.

[7] Chan, S., Pullerits, K., Keucken, A. *et al.* 2019 Bacterial release from pipe biofilm in a full-scale drinking water distribution system. *npj Biofilms Microbiom.* **5, 9 1-8.**

[8] Farrand SK, Qin Y, Oger P 2002 Quorum-sensing system of Agrobacterium plasmids: analysis and utility. In: Methods in enzymology. Elsevier. p. 452-484.

[9] Gulati P, Ghosh M 2017 Biofilm forming ability of Sphingomonas paucimobilis isolated from community drinking water systems on plumbing materials used in water distribution. *J. Wat.Health.***15** 942-954.

[10] Gusman V, Medić D, Jelesić Z, Mihajlović-Ukropina M 2012 Sphingomonas paucimobilis as a biofilm producer *Arch. Biol.Sc.***64**1327-1331.

[11] Jang H-J, Choi Y-J, Ka JO 2011 Effects of diverse water pipe materials on bacterial communities and water quality in the annular reactor *J. Microbiol. Biotechnol.***21**115-123.

[12] Kanamori H, Weber DJ, Rutala WA 2016 Healthcare outbreaks associated with a water reservoir and infection prevention strategies *Clin. Infect.Dis.***62**1423-1435.

[13] Kaur T, Ghosh M 2017 Characterization and toxicity of a phosphate-binding exobiopolymer produced by Acinetobacter haemolyticus MG606 *J. Wat. Health* **15**(1) 103-111.

[14] Khaira G, Ganguli A, Ghosh M 2014 Synthesis and evaluation of antibacterial activity of quaternized biopolymer from Klebsiella terrigena *J Appl. Microbiol.* **116** 511-518.

[15] Kreft JU, Picioreanu C, Wimpenny JW, van Loosdrecht MC 2001 Individual-based modelling of biofilms *Microbiol.***14**7 2897-2912.

[16] Kuehn M, Mehl M, Hausner M, Bungartz HJ, Wuertz S 2001 Time-resolved study of biofilm architecture and transport processes using experimental and simulation techniques: the role of EPS *Wat. Sci. Tech.***43** 143-151.

[17] Laskin A, White D 1999 Preface to special issue on Sphingomonas *J. Ind. Microbiol.Biotech.* **23** 231-235.

[18] Laspidou CS, Rittmann BE 2004 Modeling the development of biofilm density including active bacteria, inert biomass, and extracellular polymeric substances *Wat. Res.***38** 3349-3361.

[19] Lehtola MJ, Miettinen IT, Lampola T, Hirvonen A, Vartiainen T, Martikainen PJ 2005 Pipeline materials modify the effectiveness of disinfectants in drinking water distribution systems *Wat. Res.***39**1962-1971.

[20] Lehtovaara BC, Verma MS, Gu FX 2012 Multi-phase ionotropic liquid crystalline gels with controlled architecture by self-assembly of biopolymers *Carb. Poly.***87** 1881-1885.

[21] Liu Y, Tay JH 2002 The essential role of hydrodynamic shear force in the formation of biofilm and granular sludge *Wat.Res.***36**1653-1665.

[22] McClean KH, Winson MK, Fish L, Taylor A, Chhabra SR, Camara M, Daykin M, Lamb JH, Swift S, Bycroft BW 1997 Quorum sensing and Chromobacterium violaceum: exploitation of

violacein production and inhibition for the detection of N-acylhomoserine lactones *Microbiol.***143** 3703-3711.

[23] McClean KH, Winson MK, Fish L, Taylor A, Chhabra SR, Camara M, Daykin M, Lamb JH, Swift S, Bycroft BW *et al* 1997 Quorum sensing and Chromobacterium violaceum: exploitation of violacein production and inhibition for the detection of N-acylhomoserine lactones *Microbiol.***143** 3703-3711.

[24] Morohoshi T, Kato M, Fukamachi K, Kato N, Ikeda T 2008 N-acylhomoserine lactone regulates violacein production in Chromobacterium violaceum type strain ATCC 12472 *FEMS Microbio. Lett.***279** 124-130.

[25] Oie S, Oomaki M, Yorioka K, Tatsumi T, Amasaki M, Fukuda T, Hakuno H, Nagano K, Matsuda M, Hirata N 1998 Microbial contamination of 'sterile water'used in Japanese hospitals *J.Hosp. Infect.***38** 61-65.

[26] Ono M, Nikaido T, Ikeda M, Imai S, Hanada N, Tagami J, Matin K 2007 Surface properties of resin composite materials relative to biofilm formation *Den. Mat. J.* **26** 613-622.

[27] Pegoraro E, Borsato C, Dal Bello F, Stramare R, Fanin M, Palu G, Angelini C 2008 Sphingomonas paucimobilis associated with localised calf myositis *J. Neurol, Neurosurg. Psychiat.* **79** 1194-1195.

[28] Perola O, Nousiainen T, Suomalainen S, Aukee S, Kärkkäinen U-M, Kauppinen J, Ojanen T, Katila M 2002 Recurrent Sphingomonas paucimobilis-bacteraemia associated with a multi- bacterial water-borne epidemic among neutropenic patients *J. Hosp. Infect.***50**196-201.

[29]Purevdorj B, Costerton JW, Stoodley P 2002 Influence of hydrodynamics and cell signaling on the structure and behavior of Pseudomonas aeruginosa biofilms *App.Env.Microbiol.***68** 4457- 4464.

[30] Rittmann B 1982 Comparative performance of biofilm reactor types *Biotech.Bioengg.* **24**
1341-1370.

[31] Simoes M, Simões LC, Vieira MJ 2010 A review of current and emergent biofilm control strategies *LWT-Food Sc. Tech.***43** 573-583.

[32] Singer G, Besemer K, Schmitt-Kopplin P, Hödl I, Battin TJ 2010 Physical heterogeneity increases biofilm resource use and its molecular diversity in stream mesocosms *PloS one* **5**(4) 1- 11.

[33] Singh RP, Jain S, Ramarao P 2013 Surfactant-assisted dispersion of carbon nanotubes: mechanism of stabilization and biocompatibility of the surfactant *J. Nanopart.Res.* **15** 1985-1991.

[34] Teodósio J, Simões M, Melo L, Mergulhão F 2011 Flow cell hydrodynamics and their effects on E. coli biofilm formation under different nutrient conditions and turbulent flow *Biofoul.*27 1- 11.

[35] Venugopalan V, Kuehn M, Hausner M, Springael D, Wilderer P, Wuertz S 2005 Architecture of a nascent Sphingomonas sp. biofilm under varied hydrodynamic conditions *App. Env. Microbiol.*71 2677-2686.

Synthesis and Characterization of a complex mixture based on cobalt and tin and study of its application towards production of hydrogen gas from ethanol

Swapna Samala[1*],GouriSankhar Brahma[1*] and Trilochan Swain[2]

[1]Faculty of Science and Technology (IcfaiTech), the ICFAI Foundation for HigherEducation, Hyderabad, Telangana, India.
[2]P.G. Department of Chemistry, Fakir Mohan University, Nuapadhi, Balasore,Odisha, India

*Corresponding author: swapna.samala@ymail.com

Abstract. The Cobalt complex, (Co-DMET) has been synthesized from the reaction between cobalt chloride hexahydrate and dimethoxy ethane (DME) in a 1:1 molar ratio and characterized using various measurement techniques. The structural formula of this complex derived based on characterization is $[Co(DMET)_2(H_2O)_2]Cl_2.4H_2O$. Dimethoxy ethane is an O-donating ligand able to coordinate in a chelate mode to the central metal cobalt to form the complex. The bidentate ligand is used to make the complex more stable. This complex was further mixed with the catalytic amount of SnO_2 to obtain a complex mixture, (Co-DMET-Sn) and its application towards hydrogen production from ethanol at temperature 333 K was checked. Liberation of H_2 gas is found to be maximum at 15 cc of ethanol when charged with 1: 0.1 molar ratios of cobalt-DMET complex and SnO_2. This same cobalt-DMET-tin mixture was charged with various quantities of ethanol ranging from 5.0 cc to 50.0 cc at temperature 333 K. Amount of $H_2(g)$ liberated from ethanol was estimated to be 14.92 % using Gas Liquid Chromatography(GLC). The cobalt is used due to less expensive than the other metals such as Pt which is generally used for catalytic purpose. Crystallite size of the Cobalt-DMET-tin complex was found to be 23.4 nm.

Key words: $H_2(g)$, Co-DMET, Co-DMET-Tin, SnO_2, GLC, O-donating ligand, Chelate mode.

1. Introduction

Till date, energy extraction from fossil fuel is mostly through conventional techniques, although modern techniques are being employed to minimize production of environmentally harmful by-products. Conventional processes of energy harnessing create a great deal of environmental pollution due to release of greenhouse gases leading to climatic changes **[1]**. In order to overcome these drawbacks, many researchers are now working in the direction of generating cleaner fuel with extremely less by-products, while extracting energy from the fossil fuels. This prevailing scientific approach encouraged us to try out generating hydrogen fuel whose by-product is only water. We have tried to generate hydrogen gas from ethanol using cobalt-DMET-tin complex. The results thus obtained from the experiments are found to be highly interesting which encouraged us to report those in this manuscript. The H_2 gas can be produced from different sources such as fossil fuels, oils, water and other biomass etc. [**1-3**]. The production of H_2 gas from ethanol in various processes such aspartial oxidation, auto-thermal, steam reforming as well as using catalysis is nothing new to this world [**4-8**]. Byrd et. al. found that steam reforming of ethanol not only produce H_2 but also some unwanted by-products such as CO and CO_2 [**4**]. This steam reforming is an endothermic process and feasible thermodynamically at high temperatures [**9**]. Catalysts such as Ni/Al_2O_3, Ni/La_2O_3, etc. were used for steam reforming of ethanol at different temperatures, but it also produces the by-product like acetaldehyde along with desired product H_2. Fatsikotas et al. also found that the increased in reaction temperature increases the quantity of by-products such as CO, CO_2, CH_4 along with desired productH_2 [**10-11**]. The dry reforming of ethanol is also highly endothermic and produces some unwanted gas such as CO [**12-16**]. The SnO_2 embedded with K_2O or Zeolite Y catalyst was also used for the production of H_2 gas from ethanol and the conversion was about 89 % but at high temperature of about873 K [**17**]. We generated hydrogen from ethanol using 1:0.1 molar ratios of cobalt-DMET complex and SnO_2 at very low temperature and dry reforming process. The liberation of hydrogen is about 15 %.

2. Experimental

2.1.Materials

Cobalt chloride hexahydrate (Merck, ACS reagent grade, CAS: 7791-13-1, Assay 98%),1,2-Dimethoxyethane (DMET) (Anhydrous grade, Sigma Aldrich, CAS: 110-71-4, Assay 99.5%), Tin (IV) oxide (Sigma Aldrich, CAS: 18282-10-5 Assay ≥ 99.99%) and ethanol (Merck, ≥ 99.5%, CAS: 64-17-5) were used without further purification.

2.2.Physical measurements

The pH measurement was carried out using Elico LI-617 pH meter equipped with a combination of glass-Ag electrode glass Ag/AgCl (3mol dm^{-3} NaCl solution). The calibration of pH meter was carried out using standard procedure. The quantitative estimation of Cobalt and Tin were carried out with Inductively Coupled Plasma – Optical Emission Spectroscopy (ICP-OES) with iCAP – 6500 DUOICP – OES, thermo scientific system. The mixture was dissolved in 10 cc HNO_3 at 30% (v/v) and the volume was made up to 100 cc using high-performance liquid chromatography (HPLC)-grade water for elemental analysis. Electron spray ionization– Mass spectrometry (ESI-MS) was carried out by a Thermo Scientific Exactive-2011 system coupled with an Orbitrap Analyzer. JEOL JSM-7610F microscope was used for capturing Field Emission Scanning Electron Microscopy images of the complex. Gas Liquid Chromatography (GLC) was carried out by Nucon-GC 5700 equipment. The detection of H_2 gas was carried out using helium as the carrier gas and TCD-thermal conductivity detector was used. Bruker D8 Advance was used to carry out the Powder X-ray Diffraction (PXRD) which was performed at ST & IC, Cochin University of Science and Technology, Cochin Kerala. The ICP-OES, FE-SEM, ESI-MS and GLC were performed at Indian Institute of Chemical Technology (IICT) – Hyderabad.

The Cobalt-DMET-tin mixture was taken in a 500 ml. round bottom flask. The gas was generated withaddition of ethanol to the above cobalt-DMET complex along with catalytic amount of tin (IV) oxideat 333 K, without addition of water. The continuity of the gas evolution from the mixture present in reactor was persisted up to ~ 30 minutes. Initially the gas evolved very violently and then decreased towards the end. The gas volume was measured with respect to time using two inverted burettes containing double distilled water. These burettes were connected with 500 ml. reactor using tube. The evolved gas was collected in a 1-liter Tedlar sample bag with polypropylene fitting (Cat. No. 232.01; Batch: 78029). The hydrogen gas is not dissolved in distilled water at room temperature. So, the first burette was filled with distilled water. As the gases evolve, the water level of first burette decreases while the water level of second burette

IOP Conf. Series: Materials Science and Engineering 1170 (2021) 012002 doi:10.1088/1757-899X/1170/1/012002

increases. The amount of gas evolved was measured using the decrease of water level in the first burette. This volume was cross checked with the increase of water level in the second burette. No error was detected in between the difference of water volume in boththe burettes. The method of collection and measurement of gas is represented in outline as **Figure 1**. The various amounts of ethanol i.e. from 5.0 cc to 50.0 cc was used in different batches keeping the amount of cobalt-DMET-tin complex same in each batch.

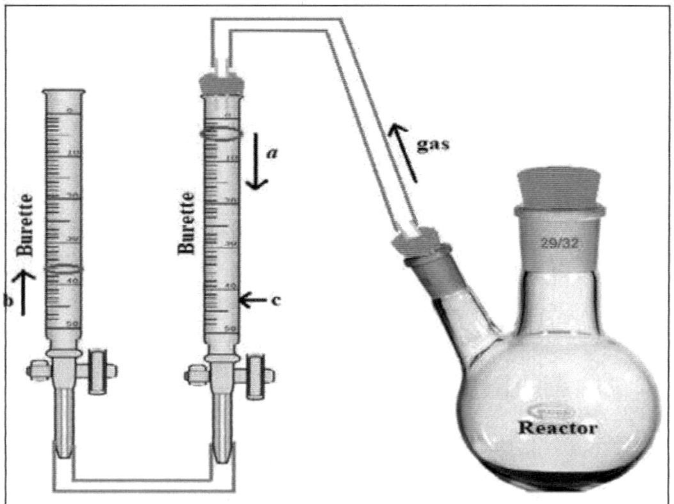

Fig. 1: The outline for collection and measurement of gas

2.3.Synthesis of Cobalt-DMET complex

Cobalt chloride hexahydrate was mixed with 1,2-dimethoxyethane (DMET) in 1:1 molar ratio. The solution turned into light blue colour. It was stirred for 30 min in room temperature in order to make it homogeneous. The temperature of the solution was increased to 363 K with stirring for two hours. Thecolor of the solution changed to dark blue. The stirring was continued for further 48 hours without heating. Then the solution was filtered, followed by evaporation to dryness. The purple colour crystals were obtained. Yield of the precipitation was 96%.

2.4.Synthesis of Cobalt-DMET-tin mixture

The physical mixture was prepared using synthesized Cobalt-DMET complex with tin (IV) Oxide in 1:
0.1 molar ratio. It was stirred for 30 minute in room temperature in order to make it homogeneous.

3. Results and Discussion

3.1.Micro-elemental analysis

3.1.1. ICP-OES analysis

The element cobalt in Co-DMET complex is quantified using ICP-OES. The experimental and theoretical values for cobalt well coincides with each other. The composition and molecular weight of complex is $[Co(DMET)_2(H_2O)_2]Cl_2.4H_2O$ and 418.14 g./mol, respectively. The calculated and experimental value of cobalt is 14.09 % and 14.22 %, respectively. Similarly, the element cobalt is quantified after the liberation of H_2 gas. The composition and molecular weight is

$[Co(DMET)_2(H_2O)_2]Cl_2.4H_2O\cdot22\ C_2H_5OH$ and 1430.14 g./mol, respectively. The calculated and experimental valueof cobalt is 4.12 % and 3.93 %, respectively. The amount of ethanol was varied in order to coincide with the experimental value of cobalt.

3.2.FTIR analysis

3.2.1. Cobalt-DMET complex

The vibrational bands of DMET ligand at 982 cm^{-1} and 938 $cm^{-1}$are completely obscured in the Co-DMET complex. This indicates that the conformation of DMET ligand changes during complex formation [**18-19**]. Fowles et al. and Snijder et al. conformed that when DMET is applied as chelating

agent then it is converted into gauche (tgt) conformation [**20-21**].This tgt conformers are highly stable in the liquid state [22]. The C-O stretching frequency band at 1131 cm^{-1} [**23**], 1155 cm^{-1} [**24**] and 1022 cm^{-1} [**25**] of DMET is shifted to 1125.05 cm^{-1} post complexation. The COC bending vibration of DMET is shifted to 472.13 cm^{-1}, which is slightly higher than that reported by Heijer and Driessen [**26**]. The bending frequency of the dimer water molecule is shifted down to 1624.07 cm^{-1} from 1640 cm^{-1} of monomer water molecule [**27**]. This indicates that there is a dimer water molecule is present in Co-DMET complex. Occurrence of the dimer is further conformed due to presence of a band at lower frequency at 601.30 cm^{-1} [**27**]. The stretching and bending mode of hydrogen bond are attributed to

670.64 cm^{-1} [**27**] instead of 700 cm^{-1}. The intra-molecular O-H stretching frequency is observed at 3405.42 cm^{-1}. The IR spectrum of cobalt-DMET complex is displayed in **Figure 2**.

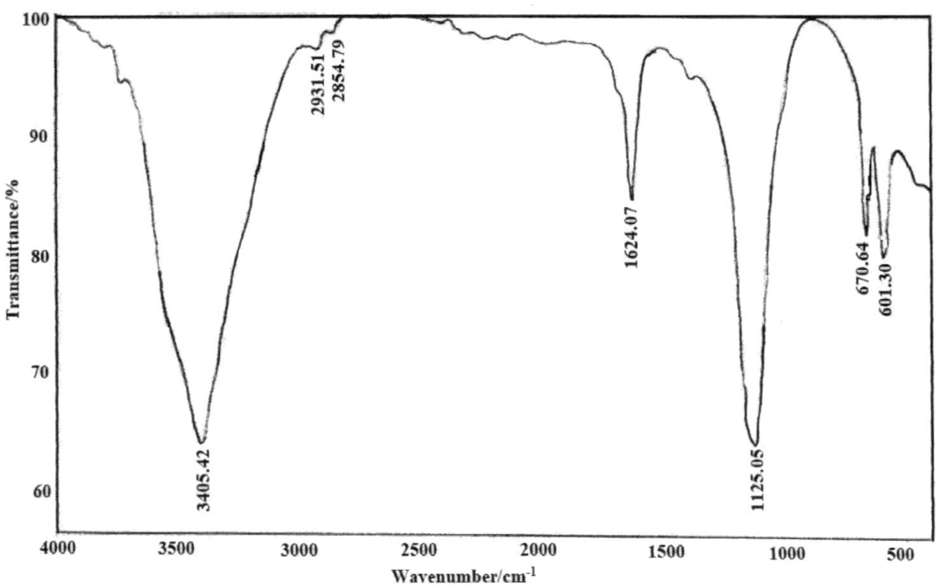

Fig. 2: IR spectrum of Co-DMET complex.

3.2.2. Cobalt-DMET-tin complex

The COC bending vibration of Cobalt-DMET-tin complex is shifted downward to 445.06 cm^{-1} as compared to Cobalt-DMET complex but slightly higher than that reported by Heijer and Driessen[**26**]. The C-O stretching frequency that appear in Cobalt-DMET complex at 1125.05 cm^{-1} is convertedto shoulder in stretching and bending mode of water band. The dimer mode of water molecule is also present in this tin associated complex at 588.89 cm^{-1} instead of 601.30 cm^{-1} of cobalt-DMET complex [**27**]. The stretching and bending mode of hydrogen bond are attributed at 652.05 cm^{-1} as compared to

670.64 cm^{-1} of cobalt-DMET complex [**27**]. The broad band at 3410.10 cm^{-1} is attributed to intermolecular stretching vibration of hydroxyl group. This is also further conformed from the strong band at 1634.06 cm^{-1} instead of 1627 cm^{-1} reported by Malik et al. [**28**]. The broad band indicates that ethanol from which H_2 is liberated form intermolecular hydrogen bonding with cobalt-DMET-tin complex at room temperature. The IR spectrum of cobalt-DMET-tin complex is displayed in **Figure 3**.

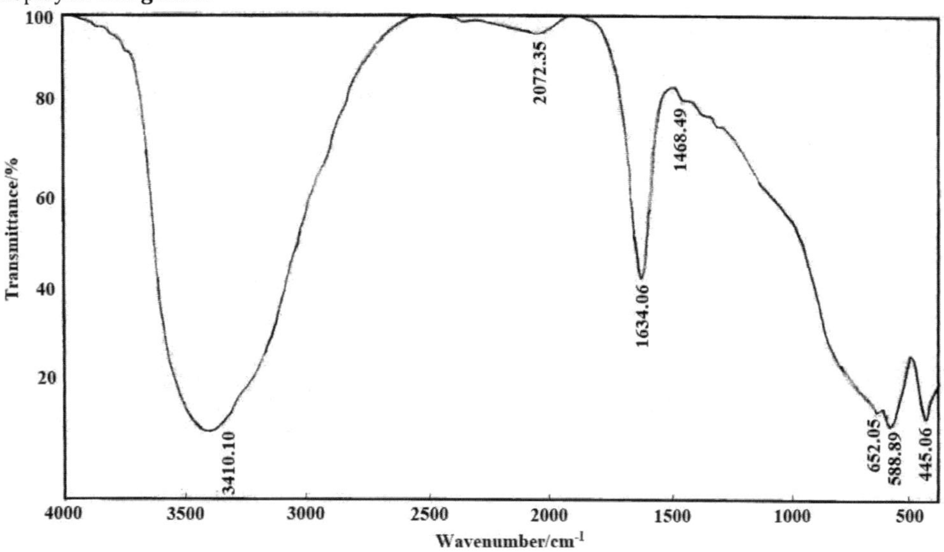

Fig. 3: IR spectrum of Co-DMET-tin complex

3.3. X-ray crystallographic analysis

The PXRD pattern of cobalt-DMET-tin complex, $[Co(DMET)_2(H_2O)_2]Cl_2.4H_2O{\cdot}22C_2H_5OH$, is displayed in **Figure 4**.

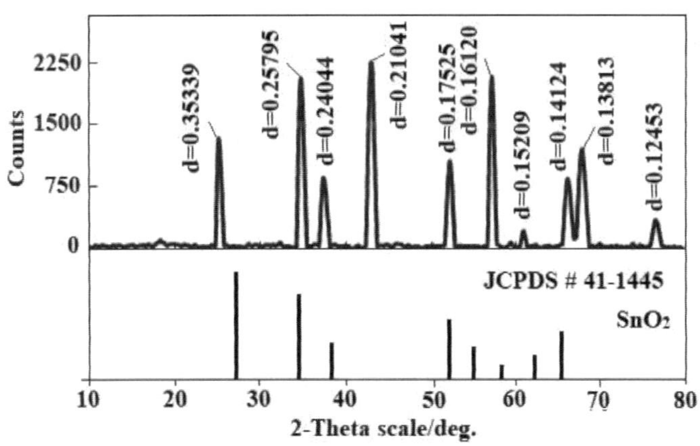

Fig. 4: PXRD pattern of cobalt-DMET-tin complex

The diffraction pattern of SnO_2 which is used as catalyst is identified using Joint Committee on Powder Diffraction Standards (JCPDS # 41-1445).

3.3.1. Crystallite size determination
The crystallite size of this cobalt-DMET-tin complex was determined using very well known the Debye-Scherrer formula. The crystallite size is 23.4 nm which is very less than the crystallite size of pure SnO_2, 60 nm [**28**]. The micro strain was also calculated using very well know formula of Stokes and Wilson. The formula is given below.

$$\varepsilon_r = \frac{\beta}{4tan\theta} \quad \text{---------} (1)$$

The micro strain at $2\boldsymbol{\theta}$ = 25.179° peak is 0.363 which is slightly less as compared to pure SnO_2, 0.385 [**28**].

3.4. FE-SEM analysis
The morphology of cobalt-DMET complex looks like the large flake of cottons that derived from the field emission scanning electron microscopy. In case of cobalt-DMET-tin complex, the morphology looks like the bunch of wet seeds. Both the images are figured in **Figure 5**. The B part of the figure 5is more fragile than the figure A. Hence, the surface area of cobalt-DMET complex increases by addition of tin (IV) oxide. As the surface area increases, the catalytic properties of cobalt-DMET-tin complex gets enhanced.

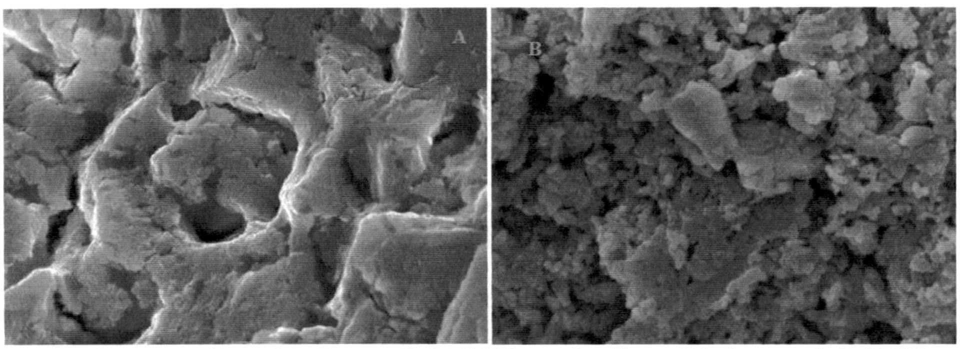

Fig. 5: SEM Images; figure A indicates cobalt-DMET complex; figure B indicates cobalt-DMET-tincomplex.

3.4.1. ESI-MS fragmentation analysis
The fragmentation pattern of cobalt-DMET complex is figured in **Figure 6**. Another figure is displayed in **Figure 7**.

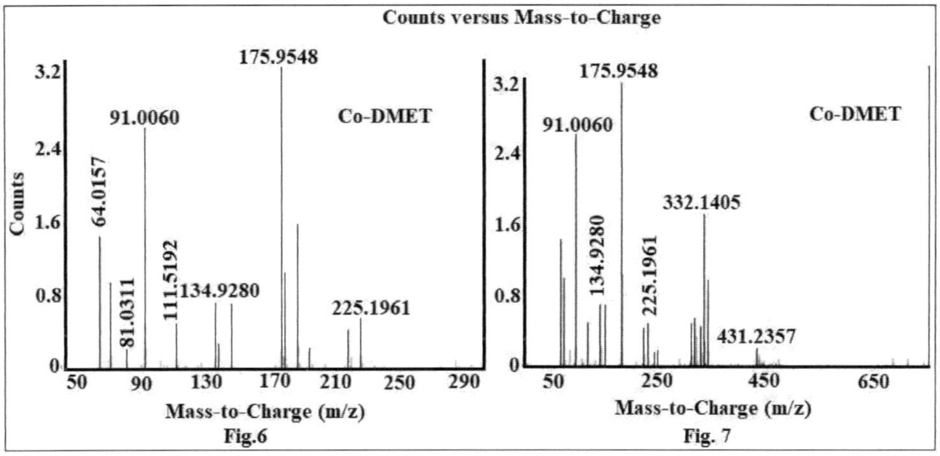

Fig. 6 & 7: The fragment pattern of cobalt-DMET complex.

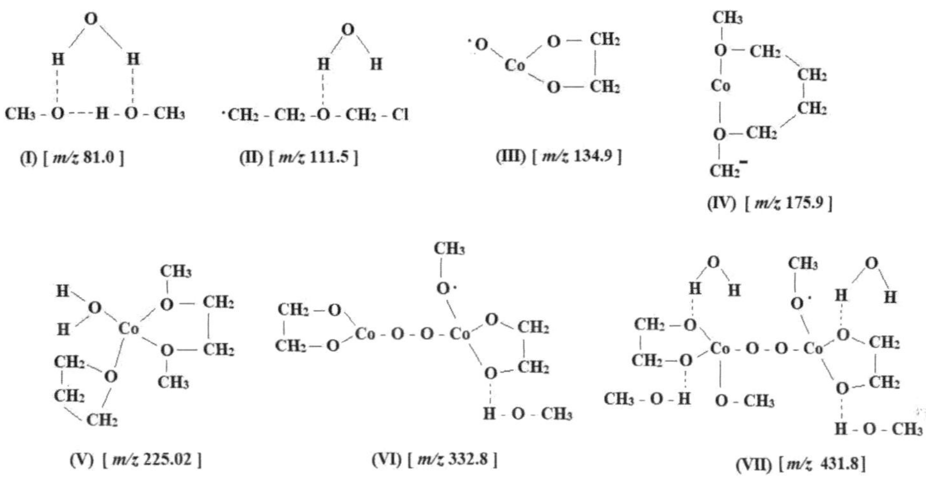

Fig. 8: Structural Fragmentation of Co-DMET Complex

For cobalt-DMET complex the important fragmentation under the experimental condition was the cleavage of DMET to form two methoxy fragments and converted into two mole of methanol *m/z* 64.These fragments of methoxy group form hydrogen bond with water having *m/z* 81. This structural fragment is displayed as (I) in **Figure 8** and corresponding empirical formula is $C_2H_9O_3$. The next fragment contain 1,2-dimthoxyetane having *m/z* 90.12 ≈ 91.0. The structural fragment having*m/z* 111.5 is displayed as (II) in **Figure 8** and its empirical formula is $C_3H_8O_2Cl\cdot$. This fragment form hydrogen bond with water. The structural fragment containing metal cobalt having *m/z* 134.9 is displayed as (III) in **Figure 8** and its empirical formula is $CoO_3C_2H\cdot$. The intense fragmentation having *m/z* 175.9 is displayed as (IV) in above mentioned figure. The empirical formula of this fragment containing cobalt metal is $O_2C_6H^-$. The fragment having *m/z* 225.02 ≈ 225.19 is displayed as (V) in above figure and the corresponding empirical formula containing metallic cobalt is$CoO_4C_7H_{18}$.Another fragment having *m/z* 332.8 ≈ 332.14 is displayed as (VI) in above figure and empirical formula containing metallic cobalt is $Co_2O_8C_6H\cdot$ .The last fragment

having m/z 431.8 ≈
431.23 and its structural unit is displayed as (VII) in **Figure 8**. The empirical formula containing cobalt is$Co_2O_{12}C_8H_{26}$

The fragmentation pattern of cobalt – DMET - tin complex is presented in **Figure 9.**

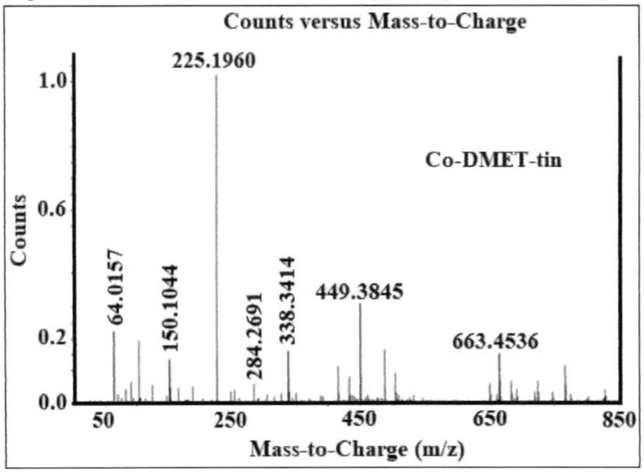

Fig. 9: ESI-MS of Co-DMET-tin complex.

For cobalt-DMET-tin complex the important fragmentation under the experimental condition was also the cleavage of DMET to form two methoxy fragments and converted into two mole of methanol m/z
64. The fragmentation having m/z 91.0 is displayed as (I) in **Figure 10**.

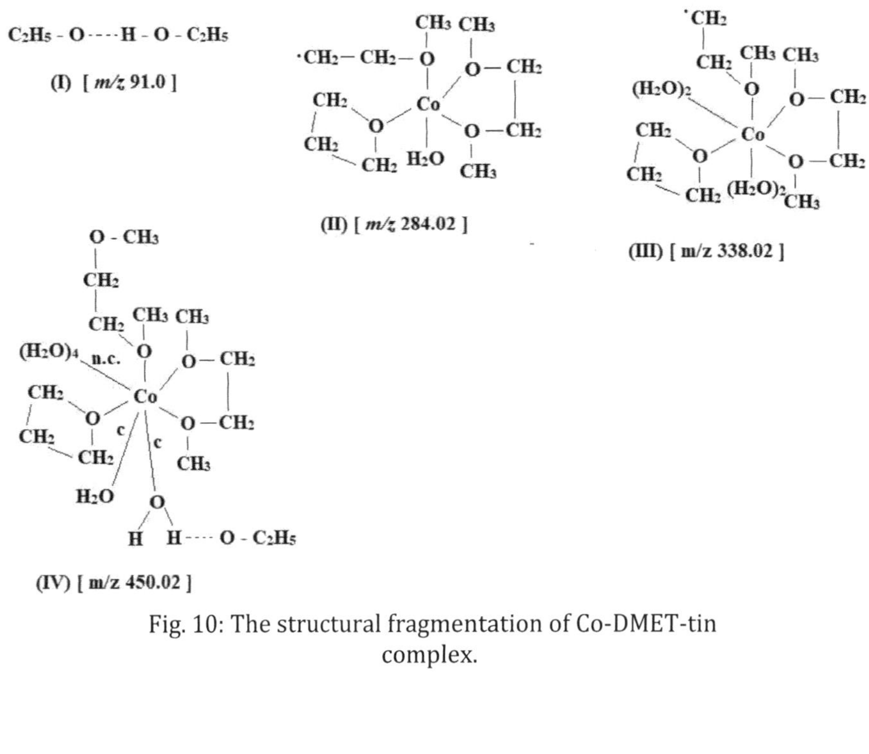

Fig. 10: The structural fragmentation of Co-DMET-tin complex.

IOP Conf. Series: Materials Science and Engineering 1170 (2021) 012002 doi:10.1088/1757-899X/1170/1/012002

In this fragmentation pattern, the ethoxide ion forms hydrogen bond with ethanol which is used in the reaction as solvent. The empirical formula for this fragmentation is $C_4H_{11}O_2$. The fragmentation having *m/z* 150.71 ≈ 150.10 belongs to SnO_2. The intense fragmentation having *m/z* 225.02 ≈ 225.19 issame as that of cobalt-DMET complex. This structural fragmentation is displayed as (V) in **Figure 8**. The next fragment having *m/z* 284.02 ≈ 284.26 and its structural unit is displayed as (II) in **Figure 10**. The empirical formula of the fragment containing cobalt is $CoO_5C_{10}H\cdot$ .This fragment when hydrated with three aqueous molecules gets converted into fragment having *m/z* 338.02 ≈ 338.34. The structuralunit is displayed as (III) in the figure. The empirical formula is $CoO_8C_{10}H\cdot$ .The above given fragment when hydrated with two aqueous molecules as well as ethoxy forming hydrogen bond with coordinated water molecule, while the methoxy form covalent bond with the free radicals of the structural unit (III). The structural unit of the above fragment is displayed as (IV) in **Figure 10**. The empirical formula containing cobalt is $CoO_{12}C_{13}H_{35}$ and having *m/z* 450.02 ≈ 449.38.

3.5. Gas Liquid Chromatography (GLC)
The retention time versus voltage graph was plotted for the detection of H_2 in Gas Liquid Chromatography (GLC). The collected gas contains 14.92 % hydrogen. Quantification of H_2 liberated with various quantity of ethanol was carried out in water. The H_2 is not absorbed in water. The chromatogram is displayed in **Figure 11**.

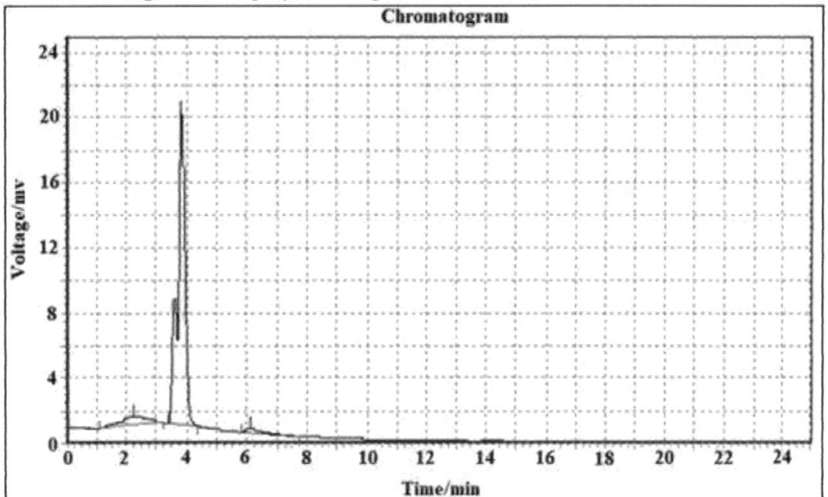

Peak No.	Rel. time	Height	Area	Conc.
1	2.208	519.957	34596.297	9.3323
2	3.837	19166.234	328590.281	88.6368
3	6.102	196.963	7529.003	2.0309
Total		19883.154	370715.581	100.0

Fig 11: GLC Chromatogram

3.5.1. Estimation of hydrogen liberation
0.01 mol. of cobalt -DMET complex is mixed with 0.001 mol. of SnO_2 for the production of H_2 from ethanol. In this experiment, the same proportion of cobalt-DMET complex and SnO_2 is mixed with various proportion of ethanol from 5.0 cc to 50.0 cc. It has been found that the maximum quantity

ofH_2 liberated is 68.8 cc at 15.0 cc of ethanol. The quantity of H_2 liberation increases with increase of addition of ethanol from 5.0 cc to 15.0 cc then decreases with increase of addition of ethanol. The quantity of H_2 liberated is 9.0 cc, 35.5 cc, 68.8 cc, 30.5 cc, 34.0 cc, 47.2 cc, 18.8 cc and 15.1 cc with respect to volume of ethanol 5.0, 10.0, 15.0, 20.0, 25.0, 30.0, 35.0, 40.0 and 50.0 cc, respectively. The abnormality of H_2 liberation after 15.0 cc of ethanol may be due to the formation of hydrogen bonding of ethoxy with the cobalt-DMET complex. The liberation of H_2 depends upon the degree of hydrogen bonding of ethoxy group. The formation of ethoxy hydrogen bonding with cobalt-DMET-tin complex is confirmed not only from the ESI-MS fragmentation pattern, **Fig 10**, but also from the FTIR analysis of cobalt-DMET-tin complex, which shows broad band at 3410 cm^{-1}.

4. Conclusion

The cobalt-DMET complex was synthesized and characterized with different measurement techniques such as FT-IR, ICP-OES, PXRD and ESI-MS. The quantity of liberation of H_2 from ethanol is measured using GLC. The 1:0.1 ratios of cobalt-DMET and tin (IV) oxide were used repeatedly for the liberation of hydrogen using various quantity of ethanol ranging from 5.0 cc to 50.0 cc at temperature 333 K. The maximum liberation of H_2 takes place when 15 cc of ethanol is charged with1: 0.1 molar ratios of cobalt-DMET complex and SnO_2 at 333 K.

Acknowledgement

SS, research scholar in Dept. of Chemistry, IcfaiTech, IFHE is thankful to the University for necessaryfinancial support. The author TS is grateful to the FM University, VyasaVihar, for granting permissionto become associated with this work.

Reference

[1] H. Balat, E. Kırtay, 2010 Int. J. Hydrogen Energy**35**, 7416.
[2] R. Kothari, D. Buddhi,R.L.Sawhney, 2008 Renew.Sust.Energ.Rev. **12**, 553.
[3] B.C.R. Ewan,R.W.K. Allen, 2005 Int. J. Hydrogen Energy**30**, 809.
[4] A.J. Byrd, K.K. Pant, R.B. Gupta, 2007 Energy Fuels**21**, 354.
[5] A. Haryanto, S. Fernando, N. Murali, S. Adhikari, 2005 Energy Fuels**19**, 2098.
[6] P.D. Vaidya,A.E. Rodrigues, 2006 Chem.Eng.*J.* **117**, 39.
[7] M. Ni, Y.C. Leung, M.K.H. Leung, 2007 *Int. J. Hydrogen Energy***32**, 3238.
[8] P.R. de la Piscina,U. Homs, 2008 *Chem.Soc. Rev.* **37**, 2459.
[9] Y. Casas-Ledón, L.E. Arteaga-Perez, M.C. Morales-Perez, L.M. Peralta-Suárez, 2012 *ISRN Thermodynamics* Article ID **672691**, 1.
[10] A.N. Fatsikostas, D.I. Kondarides,X.E. Verykios, 2001*Chem.Commun.*851.
[11] A.N. Fatsikostas, X.E. Verykios, 2004 *J.Catal.* **225**, 439.
[12] B. Bej, S. Bepari, N.C. Pradhan, S. Neogi, 2017 *Catal.Today,***291**, 58.
[13] S. Jankhah, N. Abatzoglou, F. Gitzhofer, 2008 *Int. J. Hydrogen Energy***33**, 4769.
[14] A.L. da Silva, C. de FragaMalfatti,I.L. Müller, 2009 *Int. J. Hydrogen Energy***34**, 4321.
[15] W. Wang,Y. Wang, 2009 *Int. J. Hydrogen Energy***34**, 5382.
[16] S. Liu, K. Zhang, L. Fang, Y. Li, 2008 *Energy Fuels***22**, 1365.
[17] J.S. Lee, J. Kim, M. Kang, 2011 *Bull Korean Chem.Soc.***32(6)**, 1912.
[18] R. Iwamoto, 1971 *SpectrochimActa.* **A 27**, 2385.
[19] W.L. Driessen, W.L. Groeneveld, F.W. van der Wey, 1970 *Rec. Trav.Chim.* **89**, 353.
[20] G.W.A Fowles, T.E. Lester, J.S. Wood, 1969 *J.Inorg.Nucl.Chem.* **31**, 657.
[21] R.G. Snijder, G. Zerbi, 1967 *Spectrochim.Acta.***A 23**, 391.
[22]Y. Ogawa, M. Ohta, M. Sakakibara, H. Matsuura, I. Harada, T. Shimanouchi, 1977 *Bull.Chem. Soc.* Jpn. 50, 650.
[23] H. Yoshida, T. Tanaka,H. Matsuura, 1996 *Chem.Lett.*25,637.
[24]J.P. Perdew, J.A. Chevary, S.H. Vosko, K.A. Jackson, M.R. Pederson, D.J. Singh, C. Fiolhais, 1992 *Phys. Rev. B***46**, 6671. https://doi.org/10.1103/PhysRevB.46.6671.
[25]M.J. Frisch,G.W. Trucks, H.B. Schlegel, P.M.W. Gill, B.G. Johnson, M.A. Robb, J.R. Cheeseman, T. Keith, G.A. Petersson, J.A. Montgomery, K. Raghavachari, M.A. Al-Laham,

IOP Conf. Series: Materials Science and Engineering 1170 (2021) 012002 doi:10.1088/1757-899X/1170/1/012002

V.G. Zakrzewski, J.V. Ortiz, J.B. Foresman, J. Cioslowski, B.B. Stefanov, A. Nanayakkara, M. Challacombe, C.Y. Peng,P.Y. Ayala, W. Chen, M.W. Wong, J.L. Andres, E.S. Replogle, R. Gomperts, R.L. Martin, D.J. Fox, J.S. Binkley, D.J. Defrees, J. Baker, J.P. Stewart, M. Head Gordon,C. Gonzalez, J.A. Pople,GAUSSIAN 94 (Revision D.3) 1995; Gaussian Inc.: Pittsburgh, PA.
[26] M.D. Heijer, W.L. Driessen, 1977 *Inorganica. Chimiea.Acta.***26**,221.
[27] [27]D.F. Coker,J.R. Reimers, R.O. Watts, 1982 *Aust.J. Phys.***35**, 623.
[28] R. Malik, V.K. Tomer, S. Duhan, S.P. Nehra, P.S. Rana, 2015 *EnergyEnviron. Focus.***4(4)**, 340

Protein Structure and Function Prediction of SARS-CoV 2: Prospective Antivirus Active Drug Binding Sites

Shivi Sharma [1, a)], Shefali Maheshwari[1, b)]

Department of Biotechnology, Hindustan College of Science & Technology, Mathura (Affiliated to Dr. .P.J. Abdul Kalam Technical University, Lucknow), Mathura-281122, Uttar Pradesh, India.(+91)

a)Corresponding author: shivi.sharma.hcst@sgei.org

b)Shefalisagar0315@gmail.com

Abstract: Today a newly emerged corona-virus known as SARS-CoV 2 has become a cause of global health concern and took away the lives of large number of people throughout the world. Corona-viruses are the enveloped virus with positive single stranded genome of 26.4 to 31.7 kb. Envelop of the corona-virus is made up of four structural proteins namely envelop protein (E), membrane protein (M), spike protein (S) and nucleocapsid protein (N). These four proteins are responsible for the overall shape and size of the virus (structure of virus). Envelop protein forms ion channels, membrane protein is responsible for the shape of the virus, spike protein is responsible for the entering inside the target host cell by binding to host receptor and nucleocapsid protein binds to the single stranded RNA genome of the virus forming multiple copies. We investigate the reliability and homogeneity among all the corona-virus species such as MERS CoV, Bat-CoV HKU4, Transmissible gastro-enteritis coronavirus (TGEV),Porcine epidemic diarrhea virus (PEDV),*HCoV-229E* andWhale-CoV SW1, M-CoV, Hedgehog coronavirus 1, Bulbul-CoV HKU11 etc, using a binary graph which is helpful in the findings of sequence reliability, secondary and tertiary structure model prediction using advanced model builder,we build the model of various protein/gene products by selecting them from SARS-COV 2, which further helpful in the finding of target-ligand binding for future therapeutic applications. Consequently by modeling the structure of the proteins we bring into being that envelop protein have pentameric protein lipid pores that allow ion transportation and were able to depict the active drug binding sites.

INTRODUCTION

A coronavirus is a large group of RNA viruses that causes various respiratory tract infections in humans, birds, and animals. Coronaviruses are a single-stranded positive RNA virus that belongs to the coronaviridae family. Corona is a Latin word that means a crown; as the name suggests that these viruses have crown-like projections on their surface [10].

Since 8000 B.C.E., these viruses have been around. Warm-blooded flying vertebrates (bats and birds) serve as natural reservoirs for these viruses [37]. These Coronaviruses are further divided into four sub-species which have their origin from bats and birds namely alpha-coronavirus, beta-coronavirus, gamma-coronavirus, and delta- coronavirus. The origins of alpha and beta coronaviruses are from bat and of gamma and delta from birds.

IOP Conf. Series: Materials Science and Engineering 1170 (2021) 012003 doi:10.1088/1757-899X/1170/1/012003

Coronaviruses are spherical-shaped viruses with a crown-like projection on the surface and have an average diameter of 80 to 120 nm. Coronavirus has a molecular weight of 40,000 kDa on average[10]. Coronavirus genome is of size 26.4 to 31.1 kb. 5′- leader- UTR - replicase (ORF1ab)-spike (S)-envelope (E)-membrane (M)- nucleocapsid (N) - 3′UTR-poly (A) tail is the genome organization of these crown viruses. The open reading frames 1a and 1b constitute two third of the genome and encodes polyprotein (pp1ab). 16 non-structural proteins(nsp1 to nsp16) are formed by the cleavage of replicase polyprotein. Four structural proteins envelop protein (E),membrane protein (M), nucleocapsid protein (N) and spike protein (S) are encoded by open reading frames [27]. These 4 structural proteins are responsible for the structure of the virus along with entering the host and protectingthe virus when it is outside the host. The lipid bilayer of the virus is composed of an envelope (E), spike (S), andmembrane (M) protein in the ratio 1:20:300 [27].

The first coronavirus disease was experiential in 1930 in domesticated chickens. The coronavirus responsible for this disease was known as infectious bronchitis virus (IBV) [23]. Furthermore, in 1940, two more coronavirus species, transmissible gastroenteritis virus (TGEV) [43] and hepatitis mouse virus (MHV) [45], were identified for the infection in pigs and mice, respectively. Later, in 1960, the first human coronavirus, B814, was discoveredby isolating it from a boy suffering from the common cold. In 1961, B814 Coronavirus attempted to cultivate by using the same standard techniques that have previously been used successfully for the cultivation of other common cold viruses, such as rhinoviruses and adenoviruses, but the idea failed. Later, in 1965, the virus was successfully cultivated in human embryonic trachea organ culture by serially passing through it [25]. In 1962, thecommon cold virus was isolated from a medical student and cultivate in kidney tissue culture was named 229E [17]. All these viruses, including animals and humans, had relationships with each other, but it was unknown during that time. In 1967, the structure of IBV, B814, and 229E using an electron microscope by Scottish virologists June Almeida and Tyrrell and found that these viruses were showing a relatable shape and distinctiveclub-like spikes. In the same year, the National Institution of Health isolated one more virus using the organ culture method and named it OC43 [48]. This virus was also causing cold and had a distinctive club-like spike on the surface has been observed in the electron microscope. Further, IBV showed morphological similarity with MHV, B814, 229E, and OC43; thus, this group was termed coronavirus. Later, in 2003, 2004, and 2012 other human coronaviruses came into existence. SARS-CoV in 2003, HKU1 in 2004, MERS-CoV in 2012, and currently a new coronavirus strain, SARS-CoV 2 in 2019 have been discovered [47]. Among all these viruses, SARS-CoV 2 spread among humans and become an outbreak.

At present, SARS-CoV 2, a newly emerged coronavirus, has been infecting the entire world and has become a reason for severe health concerns. This newly emerged coronavirus has taken away many lives, and no proper medication is available for the treatment of the disease. The spike protein binds to ACE2, cell surface receptor ofhost. ACE2 is also called as ACEH. It is a part of the dipeptidyl – carboxy dipeptidase angiotensin converting enzyme (ACE) family. Angiotensin 2 is converted to Ang 1–7 by ACE2 receptor. In many physiologic functions, such as the cell proliferation and hypertrophy, inflammatory response, blood pressure and the fluid balancing, ACE2 has high affinity to Type 1 and Type 2 Ang receptors. ACE2 is ubiquitously expressed in the lungs. And hence when Spike (S) protein binds to ACE 2 receptor to enter the host cell it directly targets the lung cells and causes respiratory dysfunction [50]. Cytokine release syndrome is a prominent characteristic of severe SARS- CoV-2 infections. A respiratory epithelial cell infection activates monocytes, macrophages, and dendritic cells resulting in a variety of proinflammatory cytokines, including interleukin-6, being secreted by SARS-CoV 2 (IL-6). IL-6 release causes a cascade of amplification, directly leading among other lymphocytic changes to T helper 17 (Th17). Circulation of IL-6 receptor complexes and soluble IL-6 receptor complexes indirectly activate severaltypes of cells, including endothelial cells, which cause systemically produced cytokines to cause hypotension andacute respiratory distress syndrome and this result to the

IOP Conf. Series: Materials Science and Engineering 1170 (2021) 012003 doi:10.1088/1757-899X/1170/1/012003

SARS-CoV 2 to escape from human immune system and cause the disease [51]. The risk factors for seriousness and mortality of COVID-19 is high for the persons who suffer from diabetes, hypertension, and cardiovascular diseases in particular. Vaccination is the only way to prevent the disease and stop it from spreading, for this work on the vaccine is in progress. But this virus is evolving and showing mutations and become more life-threatening.

Scientists have been discovering that this newly emerged coronavirus has become more virulent due to the occurrence of continuous mutation since November 2019. To know more about this deadly virus and its mutation,it is necessary to study its evolutionary relationship with other coronavirus species. And hence, it is required to study the structure and function prediction of the viral protein in detail.

METHODLOGY

Data retrieving and analysis

Twenty-four different coronavirus genome sequences were retrieved (as shown in TABLE 1) using NCBI for the analysis of the relationship among these different species of coronaviruses. Further, SARS-CoV 2 structuralproteins known as envelop protein, membrane protein, nucleocapsid protein and spike protein amino acid sequences (as shown in TABLE 2) were used for secondary structure prediction and tertiary structure predictionof proteins.

TABLE1: List of coronaviruses

S.NO	CORONAVIRUS	NCBI ID	SUB-TYPE
01	Avian Coronavirus	NC_001451	Gamma-CoV
02	Beluga whale coronavirus SW1	NC_010646	Gamma-CoV
03	Bovine coronavirus	NC_003045	Beta-CoV
04	Bulbul coronavirus HKU11	NC_011547	Delta-CoV
05	Feline coronavirus	NC_002306	Alpha-CoV
06	Hedgehog coronavirus 1	NC_039207	Beta-CoV
07	Human coronavirus 229E	NC_002645	Alpha-CoV
08	Human coronavirus HKU1	NC_006577	Beta-CoV
09	Human coronavirus NL63	NC_005831	Alpha-CoV
10	Human coronavirus OC43	NC_006213	Beta-CoV
11	Middle east respiratory syndrome (MERS) coronavirus	NC_019843	Beta-CoV
12	Miniopterus bat coronavirus 1	NC_010437	Alpha-CoV
13	Miniopterus bat coronavirus HKU8	NC_010438	Alpha-CoV
14	Murine coronavirus	NC_048217	Beta-CoV
15	Phipisterellus bat coronavirus HKU5	NC_009020	Beta-CoV
16	Porcine epidemic coronavirus	NC_003436	Alpha-CoV
17	Porcine coronavirus HKU15	NC_039208	Delta-CoV
18	Rhinolophus bat coronavirus HKU2	NC_009988	Alpha-CoV
19	Rousettus bat coronavirus HKU9	NC_009021	Beta-CoV
20	Scotophilus bat coronavirus 512	NC_009657	Alpha-CoV
21	Sever acute respiratory syndrome (SARS)	NC_004718	Beta-CoV
22	Sever acute respiratory syndrome	NC_045512	Beta-CoV

	coronavirus 2(SARS-CoV 2)		
23	Transmissible gastroenteritis coronavirus(TGEV)	NC_038861	Alpha-CoV
24	Tylonycteris bat coronavirus HKU4	NC_009019	Beta-CoV

TABLE2: List of structural proteins of SARS-CoV 2

S.NO	STRUCTURAL PROTEINS	ACCESSION NO.
01	Envelop protein (E)	YP_009724392
02	Membrane protein (M)	YP_009724393
03	Nucleocapsid protein (N)	YP_009724397
04	Spike protein (S)	YP_009724390

Phylogenetic analysis

Analysis of all retrieved genome sequences and the relationship among these genome sequences had performedby aligning them using MultAlin and MEGA X [41]. The phylogeny has been derived using maximum likelihood method.

Sequence analysis and Secondary structure prediction

Sequence analysis and secondary structure evaluation of the SARS-CoV 2 structural proteins have been performed to know physiochemical parameters and packing of amino acid in alpha helix, beta-sheets, and beta turns using PortParam [13] and SOPMA, [06] respectively.

Tertiary structure prediction and validation

We had used SWISS-MODEL to perform template-based modeling. To evaluate the templates of SARS-CoV 2 four structural proteins, we had individually analyzed them using BLASTp [40] against all individual structural proteins to know the percentage identity and similarity. Based on BLASTp results suitable template was used to model the tertiary structure of the protein. For the predicted models, the overall quality was validated using ERRAT [05] and ProCheck servers [26] based on the Ramachandran plot. Then the predicted models were analyzed to know about the functional properties of the structural viral proteins.

RESULTS AND DISCUSSIONS

As per the following procedure discussed above, we predict the evolutionary relationship among the different coronavirus species. And modeled the three-dimensional structures of the four structural proteins of SARS-CoV 2 (envelop protein, membrane protein, nucleocapsid protein and spike protein), the virus responsible for the ongoing pandemic worldwide. We were also able to predict the physicochemical and biochemical functions of the proteins.

```
[ 1] #NC_001451.1_Avian_infectious_bronchitis_virus_complete_genome
[ 2] #NC_039208.1_Porcine_coronavirus_HKU15_strain_HKU15-155_complete_genome
[ 3] #NC_011547.1_Bulbul_coronavirus_HKU11-934_complete_genome
[ 4] #NC_045512.2_Severe_acute_respiratory_syndrome_coronavirus_2_isolate_Wuhan-Hu-1_complete_genome
[ 5] #NC_010646.1_Beluga_Whale_coronavirus_SW1_complete_genome
[ 6] #NC_004718.3_SARS_coronavirus_Tor2_complete_genome
[ 7] #NC_006213.1_Human_coronavirus_OC43_strain_ATCC_VR-759_complete_genome
[ 8] #NC_048217.1_Murine_hepatitis_virus_strain_A59_complete_genome
[ 9] #NC_019843.3_Middle_East_respiratory_syndrome-related_coronavirus_isolate_HCoV-EMC/2012_complete_genome
[10] #NC_009021.1_Rousettus_bat_coronavirus_HKU9_complete_genome
[11] #NC_009020.1_Pipistrellus_bat_coronavirus_HKU5_complete_genome
[12] #NC_009019.1_Tylonycteris_bat_coronavirus_HKU4_complete_genome
[13] #NC_006577.2_Human_coronavirus_HKU1_complete_genome
[14] #NC_039207.1_Betacoronavirus_Erinaceus/VMC/DEU/2012_isolate_ErinaceusCoV/2012-174/GER/2012_complete_genome
[15] #NC_003045.1_Bovine_coronavirus_isolate_BCoV-ENT_complete_genome
[16] #NC_038861.1_Transmissible_gastroenteritis_virus_complete_genome_genomic_RNA
[17] #NC_009988.1_Bat_coronavirus_HKU2_complete_genome
[18] #NC_003436.1_Porcine_epidemic_diarrhea_virus_complete_genome
[19] #NC_005831.2_Human_Coronavirus_NL63_complete_genome
[20] #NC_002645.1_Human_coronavirus_229E_complete_genome
[21] #NC_002306.3_Feline_infectious_peritonitis_virus_complete_genome
[22] #NC_010437.1_Bat_coronavirus_1A_complete_genome
[23] #NC_010438.1_Bat_coronavirus_HKU8_complete_genome
[24] #NC_009657.1_Scotophilus_bat_coronavirus_512_complete_genome
```

[	1	2	3	4	5	6	7	8	9	10	11	12	13	14	15	16	17	18	19	20	21	22	23	24]
[1]																								
[2]	16.46																							
[3]	8.10	15.32																						
[4]	16.01	8.41	16.68																					
[5]	15.40	8.52	15.35	8.17																				
[6]	9.40	11.76	8.30	14.32	14.48																			
[7]	13.78	15.70	12.14	16.09	14.43	16.07																		
[8]	8.71	14.36	9.02	14.66	15.47	8.24	14.60																	
[9]	15.00	16.64	11.88	15.87	16.13	15.40	7.43	14.79																
[10]	7.59	15.43	8.29	14.30	16.19	10.28	14.78	8.44	14.02															
[11]	11.09	16.40	15.13	13.93	15.46	14.28	7.37	14.03	7.88	15.39														
[12]	14.70	15.77	14.49	14.93	15.11	15.12	7.53	14.96	7.48	14.20	7.89													
[13]	14.98	8.97	16.42	7.26	9.27	16.50	14.36	13.70	14.01	15.62	11.62	13.79												
[14]	7.76	15.08	7.65	14.20	15.30	8.05	14.79	5.41	15.58	9.20	12.10	14.39	15.44											
[15]	8.75	11.63	8.18	15.23	13.14	7.63	18.94	7.45	14.39	10.74	16.31	11.40	14.23	7.06										
[16]	14.51	16.02	14.35	15.16	15.32	15.30	8.17	15.53	9.16	10.35	7.25	7.95	13.93	11.45	14.81									
[17]	11.49	16.87	15.64	15.38	16.62	15.47	7.17	14.11	7.35	11.16	7.17	7.28	15.19	13.93	16.42	6.96								
[18]	14.11	16.33	14.29	16.49	16.92	12.19	7.40	14.08	6.77	14.61	8.49	8.42	14.44	14.59	15.92	6.40	6.00							
[19]	16.40	7.18	16.35	7.92	8.01	15.75	15.03	15.00	13.52	15.09	14.79	10.80	7.24	15.77	10.28	14.86	16.21	15.72						
[20]	14.91	7.85	14.99	7.69	7.41	15.49	16.37	15.71	14.78	15.51	16.33	15.78	7.49	16.61	14.85	15.89	16.18	15.17	7.25					
[21]	15.22	7.92	15.90	7.87	8.75	14.02	16.08	15.31	14.68	15.09	15.70	15.66	9.41	14.62	14.56	16.44	16.97	15.98	6.69	7.70				
[22]	17.05	7.69	16.32	7.46	7.26	16.30	14.81	16.21	15.43	16.75	16.36	16.01	7.40	15.88	13.39	15.53	15.85	16.21	7.24	6.97	8.55			
[23]	11.85	17.00	10.35	16.78	16.89	16.16	7.32	14.04	7.14	11.95	7.80	7.03	14.58	11.07	16.32	6.98	6.36	6.44	15.69	15.76	16.39	15.71		
[24]	10.60	16.78	11.08	14.60	15.79	16.04	8.49	15.63	6.85	13.88	7.45	7.61	14.25	14.63	15.26	7.61	6.13	7.51	14.74	16.00	16.07	15.12	7.47	

FIGURE 1. Scoring Matrix based alignment of the above genome sequences.

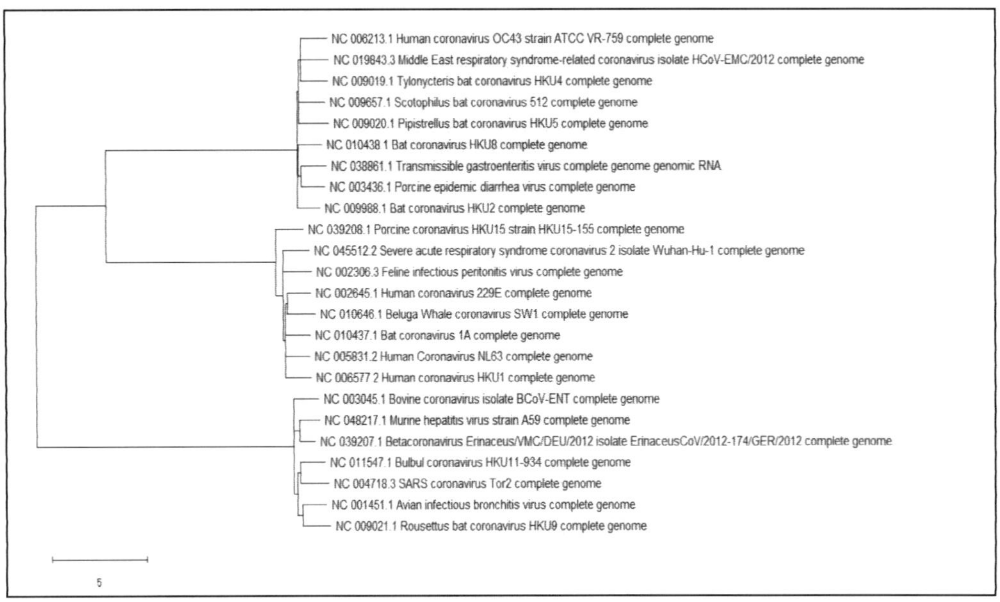

FIGURE 2. Phylogenetic tree for different coronavirus species

Envelop Protein

This protein is a minor structural protein of SARS-CoV 2, made up of 75 amino acid residues. The molecular weight of this protein is 8365.04 Daltons. The physiochemical prediction shows that envelop protein is rich in leucine (18.7%), valine (17.3%), and serine (10.7%) (FIGURE 3). The secondary structure analysis shows that the amino acid of enveloping protein [E] has 44% alpha helixes, 9% beta turns, 26.67% extended strands, and 20% random coil arrangement (FIGURE 4).

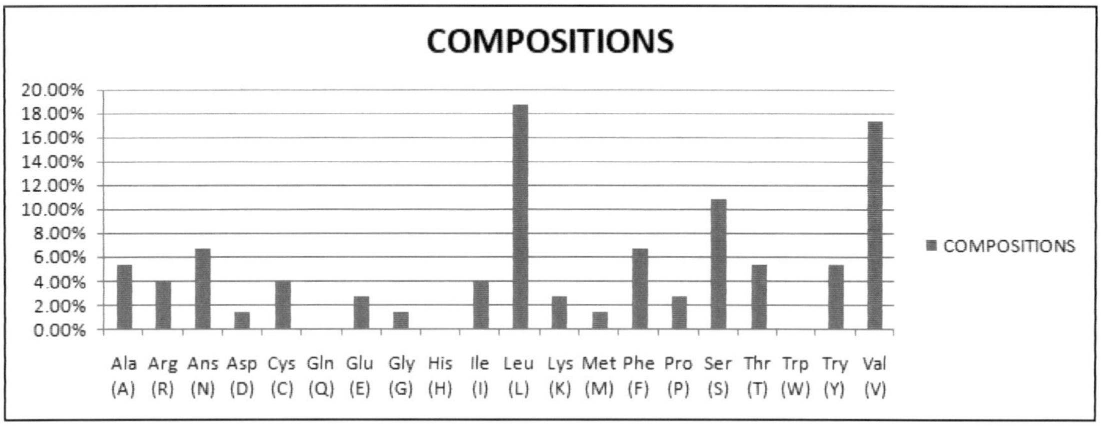

FIGURE 3. SARS-CoV 2 envelop protein amino acid compositions.

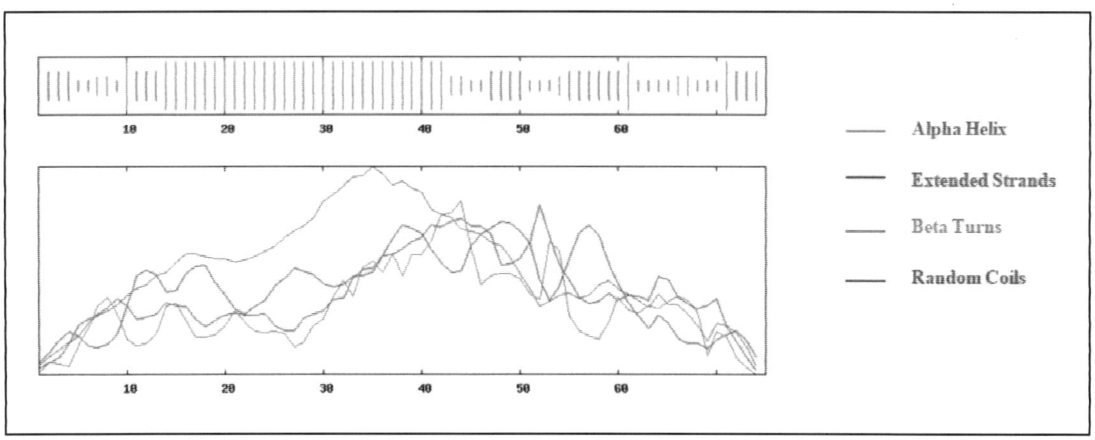

FIGURE 4. Secondary structure prediction for Envelop protein

During the modeling of the envelope protein, four templates had generated. After performing the BLASTp template used for modeling showed 91.38% percentage identity and 0.55 sequence similarities, which was an envelope small membrane proteins of conserved Golgi complex targeting signal in coronavirus envelope protein.The verified structure has a QMEAN score of -3.40 and a global score of 0.39. ProCheck verification revealed that 92.6% of residues are in the favored region (A, B, L) and 94.64% favored by the Ramachandran plot (FIGURE 5). Structure verification by ERRAT reveals the average quality of the model with an overall quality factor of 68%. The tertiary structure of envelope protein results in the pentameric ion channel and Golgi complex of protein which targeting the signals inside the SARS-CoV 2 envelope protein.

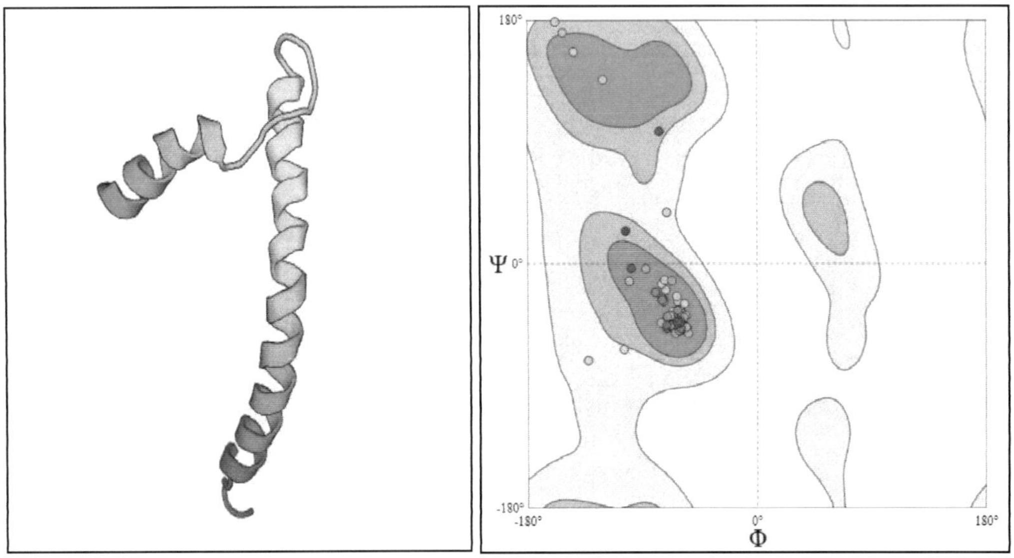

FIGURE 5. Envelop protein model and favored Ramachandran Plot.

Membrane Protein

This protein is a major structural protein that provides the shape and maintains the structure of SARS-CoV 2, made up of 222 amino acid residues. The molecular weight of this protein is 25146.62 Daltons. The physiochemical prediction shows that the membrane protein is rich in leucine (15.8%) (FIGURE 6). The secondary structure analysis shows that the amino acids of membrane protein have 34.68% alpha-helix, 6.76% beta turns, 21.17% extended strands, and 37.39% is random coil arrangement (FIGURE 7).

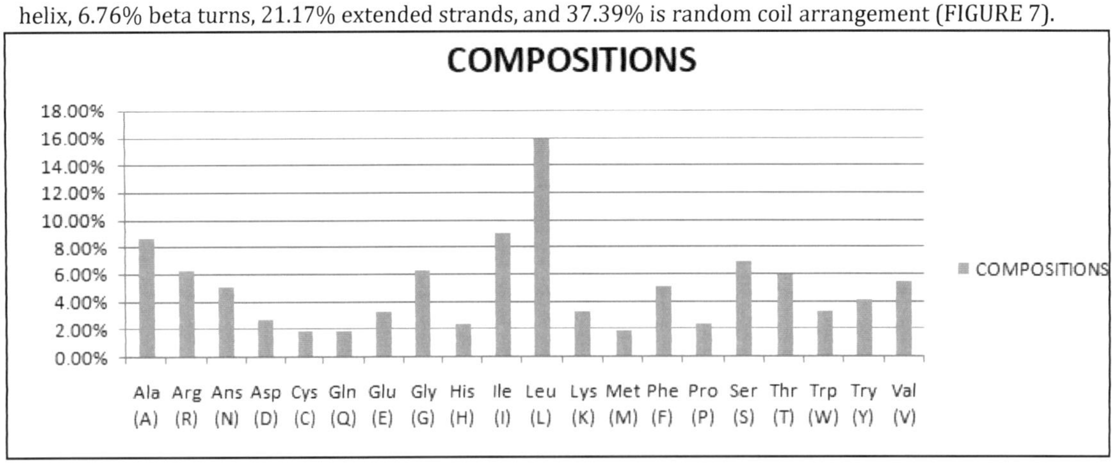

FIGURE 6. SARS-CoV 2 membrane protein amino acid compositions.

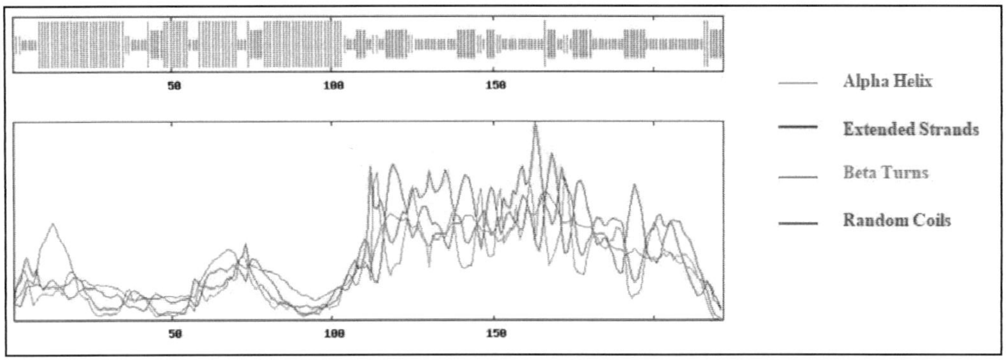

FIGURE 7. Secondary structure prediction for Membrane protein

During the modeling of the membrane protein, 14 templates had generated. After performing the BLASTp template used for modeling showed 15.38% percentage identity and 0.28 sequence similarities. The template wasan E.coli core signaling unit carrying receptor mutation. The verified structure has a QMEAN score of -2.48 anda global score of 0.3. ProCheck verification revealed that 83.3% of residues are in the favored region (A, B, L) and 93.88% favored by the Ramachandran plot (FIGURE 8). ERRAT reveals the average quality of the model with an overall quality factor of 57.14%. The tertiary structure of membrane protein results in the triple helix bundle that forms a 3-trans membrane domain in SARC-CoV 2.

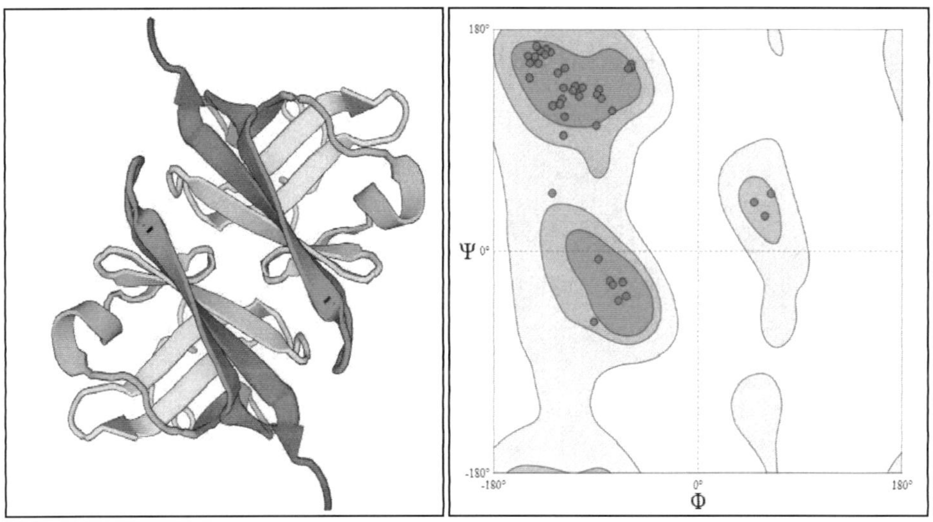

FIGURE 8. Membrane protein model and favored Ramachandran Plot

Nucleocapsid Protein

This protein binds to the single-stranded RNA genome forming multiple copies in SARS-CoV 2, made up of 419 amino acid residues. The molecular weight of this protein is 45625.70Daltons. The physiochemical prediction shows that the nucleocapsid protein is rich in glycine (10.3%) (FIGURE 9). The secondary structure analysisshows that the amino acids of nucleocapsid protein have 21.24% alpha-helix, 6.92% beta turns, 16.71% extendedstrands, and 55.13% random coil arrangement (FIGURE 10).

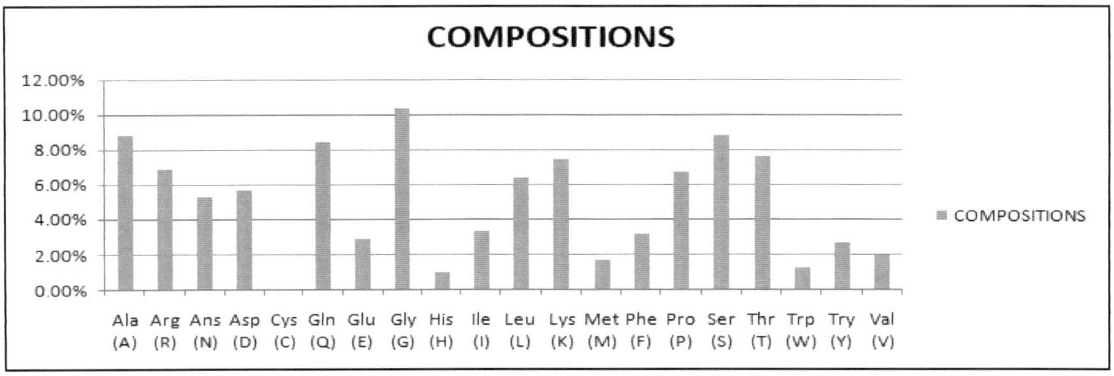

FIGURE 9. SARS-CoV 2 nucleocapsid protein amino acid compositions.

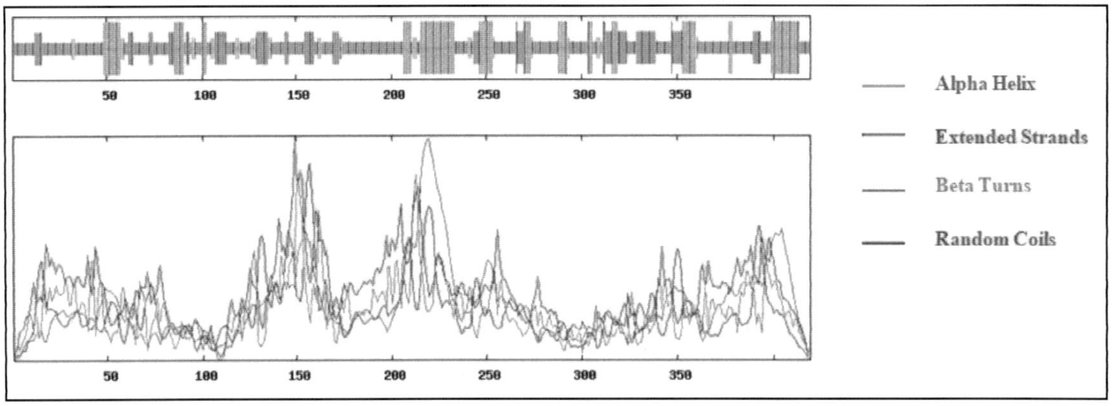

FIGURE 10. Secondary structure prediction for Nucleocapsid protein

During the modeling of the nucleocapsid protein, 36 templates had generated. After performing the BLASTp template used for modeling showed 92.06% percentage identity and 0.61 sequence similarities, which was an RNA binding domain of SARS-CoV nucleocapsid. The verified structure has a QMEAN score of 1.06 and a global score of 0.16. ProCheck verification revealed that 89.4% of residues are in the favored region (A, B, L) and 99.19% favored by the Ramachandran plot (FIGURE 11). ERRAT reveals a good quality of the model with an overall quality factor of 89.56%. The tertiary structure shows RNA binding domains present on the nucleocapsid protein of SARS-CoV 2.

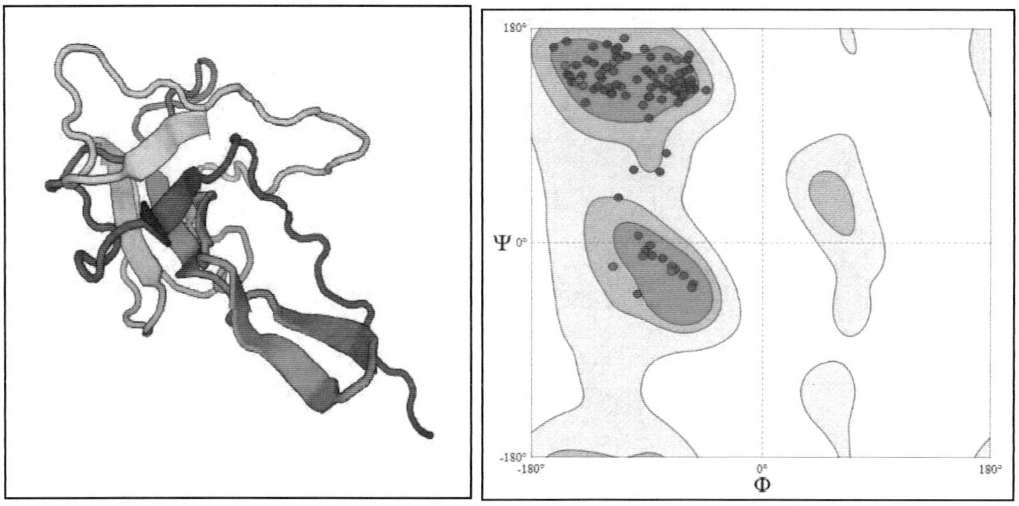

FIGURE 11. Nucleocapsid protein model and favored Ramachandran Plot

Spike Protein

This protein binds to the host receptor for entering the host cell, made up of 1273 amino acid residues. The molecular weight of this protein is 141178.47Daltons. The physiochemical prediction shows that the spike proteinis rich in leucine (8.5%) (FIGURE 12). The secondary structure analysis shows that the amino acids of spike protein have 28.59% alpha-helix, 3.38% beta turns, 23.25% extended strands, and 44.78% is random coil arrangement (FIGURE 13).

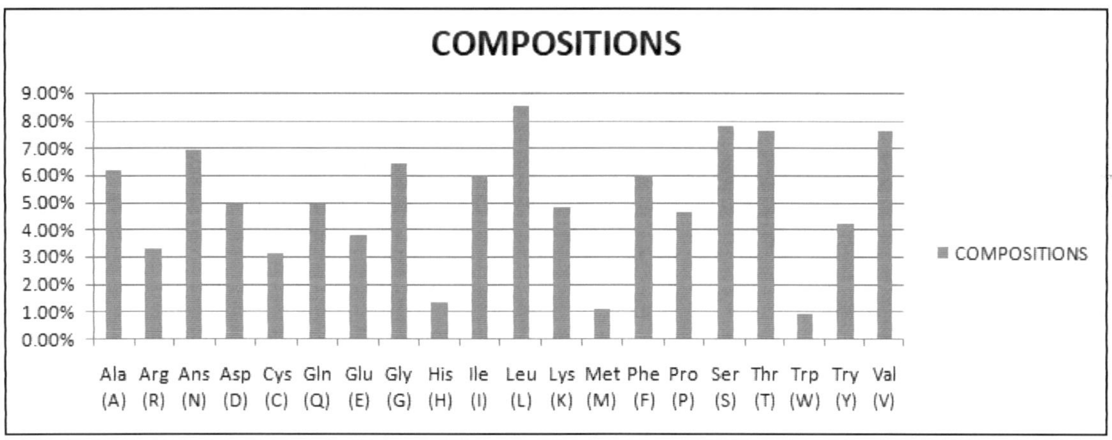

FIGURE 12. SARS-CoV 2 spike protein amino acid compositions.

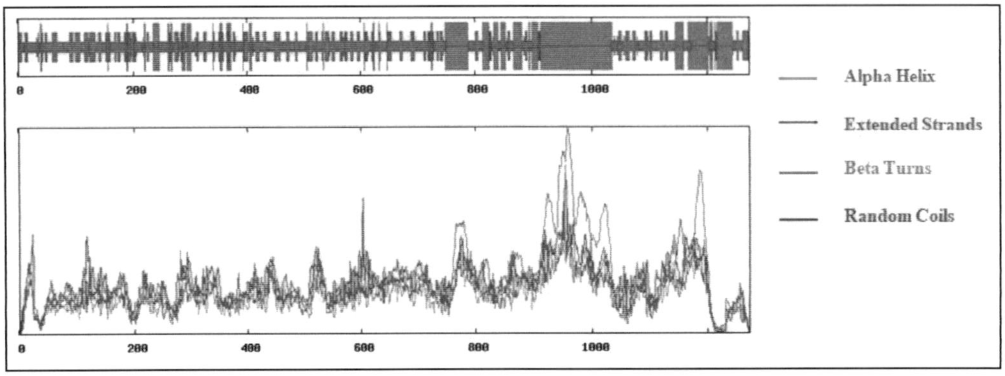

FIGURE 13. Secondary structure prediction for Spike protein

During modeling of spike protein, 34 templates had generated. After performing the BLASTp template used for modeling showed 76.79% percentage identity and 0.55 sequence similarities were selected, which was trypsin cleaved low pH treated SARA-CoV spike glycoprotein. The verified structure has a QMEAN score of -3.72 and a global score of 0.60. ProCheck verification revealed that 80% of residues are in the favored region (A, B, L) and 87.11% favored by the Ramachandran plot (FIGURE 14). ERRAT reveals a good quality of the model with an overall quality factor of 84.53%. The tertiary structure of membrane protein results that spike protein has two functional subunits S1 and S2. S1 is globule and has receptor binding domains (RBD) that bind to the target hostreceptor and enters the cell. While S2 is the stem that is for membrane fusion (Fig14)

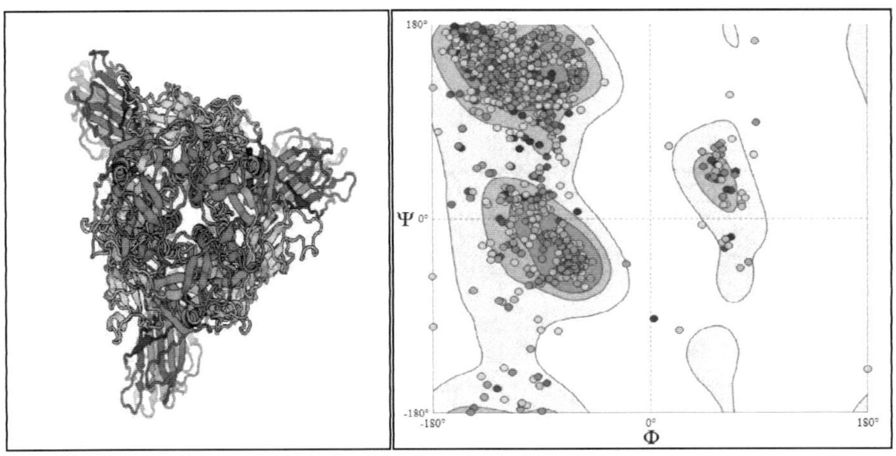

FIGURE 14. Spike protein model and favored Ramachandran Plot.

CONCLUSION

The evolutionary relation generated by the phylogenetic tree on the basis of its scoring matrix using maximum likelihood method, concludes that the four genera, α, β, γ, and δ coronaviruses are likely to occur in relationship with each other. For many years these coronaviruses are emerging and evolving. Their roots are from avian coronaviruses. The new classification system shows of bat coronaviruses are dominant in α-coronavirus sub- species. While in β, γ, and δ-coronaviruses, bird coronaviruses are dominant. The enormous variety of coronaviruses originated from bats and birds make an excellent gene pool of alpha, beta, gamma and delta-coronaviruses [37].

Analyzing the SARS-CoV 2 structural proteins shows that the envelope protein (E) is 8.4 to 12kDa in size and consists of 76 amino acid residues [30]. This protein is an integral protein that is embedded into the lipid layer. The transmembrane domain and extra membrane C-terminal domain are the two domains of the E protein. This protein is vital for creating a viral assembly. There are almost 20 E proteins in the virus and forms an alpha-helical structure with a single alpha-helical transmembrane domain and forms a five-molecular ion channel in lipid bilayer[12]. Membrane protein (M) is a 7.8nm thick structural protein, consists of 222 amino acid residues. This proteinis responsible for providing the overall shape of the virus [24]. M protein has three domains. One is N-terminal ectodomain, the second one is triply spanning trans-membrane domain, and the third one is C-terminal endodomain [04]. Nucleocapsid protein (N) is 43-50kDa in size, consists of 419 amino acid residues. Its functionis the packaging of approx 30kb, 5′-capped positive-strand viral genome of single stranded RNA molecule into aribonucleoprotein (RNP) complex known as capsid. This protein is highly phosphorylated, due to which structuralchanges take place in the virus and enhances the affinity of SARS-CoV 2 [07]. Spike protein (S) is a 1273 aminoacid residue protein. That is responsible for the corona's halo-like surface. There are 20nm-sized 74 surface spikecomposed of trimmer of S protein on the virus surface approximately [36]. There are two sub-units of spike proteinS1 and S2 subunits. S1 subunit contributes as head of the protein and act as a receptor binding domain. S2 subunitcontributes to stem of the protein which harbors the spike on the virus envelope. In the functionally active SARS-CoV 2 virus, three S1 subunits get attached with two S2 subunits. When the virus binds to the receptor of the host,the subunit complexes split into individual subunits [14]. Hence this study helps to develop a potential candidatefor drug designing to fight against the viral infection.

ACKNOWLEDGEMENT
We acknowledge Hindustan College of Science and Technology, Farah Matura to provide a workplace in theirBioinformatics lab.

REFRENCES

[1] A.Tabibzadeh, M.Esghaei, S.Soltani, P.Yousefi, M.Taherizadeh, F.S.Tameshkel, M.Golahdooz, M.Hadi "Evolutionary study of COVID-19, sever acute respiratory syndrome coronavirus 2 (SARS-CoV 2) as an emerging coronavirus: phylogenetic analysis and litrature review." Vet Med Sci. 2020;00:1–13; DOI: 10.1002/vms3.394

[2] A.A.T Naqvia , K.Fatimab , T. Mohammada , U. Fatimac , I. K. Singhd , A. Singhe , S.M. Atiff , G.Hariprasadg, G.M. Hasanh , M.I Hassan "Insights into SARS-CoV-2 genome, structure, evolution, pathogenesis and therapies: Structural genomics approach" BBA - Molecular Basis of Disease 2020 1866:165878; https://doi.org/10.1016/j.bbadis.2020.165878

[3] A.Sapkota, "Structure and Genome of SARS-CoV-2 (COVID -19) with diagram" April 5, 2020. Microbialnotes. https://microbenotes.com/structure-and-genome-of-sars-cov-2/

[4] Benjamin W. Neuman,a, Gabriella Kiss,a Andreas H. Kunding,d,f David Bhella,i M. FazilBaksh,b Stephen Connelly,c Ben Droese,c Joseph P. Klaus,l Shinji Makino,h Stanley G. Sawicki,j Stuart G. Siddell,kDimitrios
G. Stamou,d,e,f Ian A. Wilson,c Peter Kuhn,c and Michael J. Buchmeierg, "A structural

analysis of M proteinin coronavirus assembly and morphology" J Struct Biol. 2011 174(1): 11–22.doi: 10.1016/j.jsb.2010.11.021

[5] C Colovos , T O Yeates "Verification of protein structures: patterns of nonbonded atomic interactions" Protein Sci 1993 Sep;2(9):1511-9. doi: 10.1002/pro.5560020916

[6] C. Geourjon, G. Deléage "SOPMA: significant improvements in protein secondary structure prediction by consensus prediction from multiple alignments" Bioinformatics 1995 11: 6, 681- 684, https://doi.org/10.1093/bioinformatics/11.6.681

[7] C.K.Changa, M.H.Houb, C.F.Changc, C.D.Hsiaod, T.H.Huangace, "The SARS coronavirus nucleocapsid protein – Forms and functions" Antiviral Research 2014 (103) 39-50; https://doi.org/10.1016/j.antiviral.2013.12.009

[8] C.Baruah, P.Devi, D.Sharma "Sequence analysis and structure prediction of SARS-CoV 2 acessory proteins 9b and ORF14: evolutionary analysis indicates close relatedness to bat coronavirus." BioMed Research International Volume (2020); https://doi.org/10.1155/2020/7234961

[9] C.Wu, Y.Liu, Y.Yang, P.Zhang , W.Zhong , Y.Wang , Q.Wang , Y.Xu , M.Li , X.Li, M.Zheng, L.Chen, H.Li "Analysis of therapeutic targets for SARS-CoV-2 and discovery of potential drugs by computational methods." Acta Pharmaceutica Sinica B (2020) 10(5):766e788; https://doi.org/10.1016/j.apsb.2020.02.008

[10] C. S. Goldsmith, K. M. Tatti, T. G. Ksiazek, P. E. Rollin, J.A. Comer, W. W. Lee, P. A. Rota, B.Bankamp, William, J. Bellini, and S. R. Zaki, "Ultrastructural Characterization of SARS Coronavirus" Emerging Infectious Disease 2004. 10(2): 320–326. doi: 10.3201/eid1002.030913

[11] D. E. Gordon, G. M. Jang, N. J. Krogan "A SARS-CoV-2 protein interaction map reveals targets for drug repurposing" Nature 2020 (583)459–468; https://doi.org/10.1038/s41586-020-2286-9

[12] D. Schoeman and B. C. Fielding, "Coronavirus envelope protein: current knowledge" Schoeman and Fielding Virology Journal (2019) 16:69 https://doi.org/10.1186/s12985-019-1182-0

[13] E. Gasteiger, C. Hoogland, A. Gattiker et al., "Protein identification and analysis tools on the ExPASy server," in The Proteomics Protocols Handbook, pp. 571–607, Humana Press, 200

[14] E. A J Alsaadi, I.M Jones, "Membrane binding proteins of coronaviruses" Future Virol. 2019 14(4): 275–286;doi: 10.2217/fvl-2018-0144

[15] F. Hufsky, K. Lamkiewicz, A. Almeida, A.Aouacheria, C. Arighi, A. Bateman, J. Baumbach, N. Beerenwinkel, C. Brandt, M. Cacciabue, S. Chuguransky, O. Drechsel, R. D. Finn, A.Fritz, S. Fuchs, et al. "Computational strategies to combat COVID-19: useful tools to accelerate SARS-CoV-2 and coronavirus research" Briefings in Bioinformatics, 2020, 1–22; doi: 10.1093/bib/bbaa232

[16] F. K. Yoshimoto "The Proteins of Severe Acute Respiratory Syndrome Coronavirus-2 (SARS CoV-2 or n-COV19), the Cause of COVID-19" The Protein Journal (2020) 39:198–216 https://doi.org/10.1007/s10930-020-09901-4

[17] Geller C, Varbanov M, Duval RE, "Human coronaviruses: insights into environmental resistance and its influence on the development of new antiseptic strategies"

Viruses 2012, 4(11), 3044- 3068; https://doi.org/10.3390/v4113044

[18] Hamre D, Procknow JJ, "A new virus isolated from the human respiratory tract" Society for Experimental Biology and Medicine, 121(1), 190-193; https://doi.org/10.3181/00379727-121-30734

[19] J.A. Jaimes, N.M. André, J.S.Chappie, J.K.Millet, G.R.Whittaker "Phylogenetic Analysis and Structural Modeling of SARS-CoV-2 Spike Protein Reveals an Evolutionary Distinct and Proteolytically Sensitive Activation Loop" Journal of Molecular Biology (2020) 432, 3309–3325; https://doi.org/10.1016/j.jmb.2020.04.009

[20] J.Xu , S.Zhao , T.Teng , A.E.Abdalla , W.Zhu, L.Xie, Y.Wang, X.Guo "Systematic Comparison of Two Animal-to-Human Transmitted Human Coronaviruses: SARS-CoV-2 and SARS-CoV." Viruses 2020, 12, 244; doi:10.3390/v12020244

[21] J.F.W Chan, K.H.Kok, Z.Z.Hin Chu, K.K-Wang To, S. Yuan, K.K Yuen, "Genomic characterization of the 2019 novel human-pathogenic coronavirus isolated from a patient with atypical pneumonia after visiting Wuhan" January 28, 2020 Emerging Microbes & Infections. 10.1080/22221751.2020.1719902

[22] J.O. Wertheim, D. K. W. Chu, J. S. M. Peiris, S. L. K. Pond, L. L. M. Poon, "A Case for the Ancient Origin of Coronaviruses" Journal of Virology2013, 7039–7045; doi:10.1128/JVI.03273-12

[23] Kahn, J. S. McIntosh, Kenneth, "History and Recent Advances in Coronavirus Discovery" The Pediatric Infectious Disease Journal: 2005 24(11) S223-S227; doi: 10.1097/01.inf.0000188166.17324.60

[24] K. Lalchhandama, "The chronicles of coronaviruses: the electron microscope, the doughnut, and the spike" June 30, 2020 Journal of the Mizo Academy of Sciences https://doi.org/10.33493/scivis.20.02.03

[25] Knapp, Alex. "The Secret HistoryOf The First Coronavirus". Forbes. https://www.forbes.com/sites/alexknapp/2020/04/11/the-secret-history-of-the-first-coronavirus- 229e/#300a62da71d6

[26] Laskowski R A, MacArthur M W, Moss D S, Thornton J M (1993). PROCHECK - a program to check thestereochemical quality of protein structures. J. App. Cryst., 26, 283-291

[27] L. F. García, "Immune Response, Inflammation, and the Clinical Spectrum of COVID-19" June 16, 2020frontiers in immunology. https://doi.org/10.3389/fimmu.2020.01441

[28] M. Sarkar, S. Saha "Structural insight into the role of novel SARS-CoV-2 E protein: A potential target for vaccine development and other therapeutic strategies" PLoS ONE 15(8): e0237300.https://doi.org/10.1371/journal.pone.0237300

[29] M. Bianchi, D.Benvenut , M. Giovanetti , S. Angeletti , M. Ciccozzi, and S. Pascarella "Sars-CoV-2 Envelope and Membrane Proteins: Structural Differences Linked to Virus Characteristics?" Hindawi BioMed Research International 2020, 6, 4389089; https://doi.org/10.1155/2020/4389089

[30] M.M.C.Lai,D.Cavanagh, "The Molecular Biology of Coronaviruses" Advance in virus research.https://doi.org/10.1016/S0065-3527(08)60286-9

[31] N.Kaur a, R.Singh, Z.Dar, R.K.Bijarnia, N.Dhingra, T.Kaur "Genetic comparison among various coronavirus strains for the identification of potential vaccine targets of SARS-CoV2." Elsevier B.V. (2020)1567-1348; https://doi.org/10.1016/j.meegid.2020.104490

[32] N. Decaro " Betacoronavirus" , The Springer Index of Viruses, DOI 10.1007/978-0-387-95919-1

[33] N. Decaro "Alphacoronavirus" The Springer Index of Viruses, DOI 10.1007/978-0-387-95919-1

[34] N. Decaro "Gammacoronavirus" The Springer Index of Viruses, DOI 10.1007/978-0-387-95919-1

[35] N. K. Dutta, K. Mazumdar, J. T. Gordyc "The Nucleocapsid Protein of SARS–CoV-2: a Target for Vaccine Development" Journal of Virology 2020, 94:3 https://doi.org/10.1128/JVI.00647-20.

[36] P.S.Masters, "The Molecular Biology of Coronaviruses" Advance in virus research https://doi.org/10.1016/S0065-3527(06)66005-3

[37] P.C. Y. Woo, S.K. P. Lau, C. S. F. Lam, C. Y. Lau, A. K. L. Tsang, J.H. N. Lau, R. Bai, J. L. L. Teng, C.C.

C. Tsang, M. Wang, B.J. Zheng, K.H. Chan, and K.Y. Yuen "Discovery of Seven Novel Mammalian and Avian Coronaviruses in the Genus Deltacoronavirus Supports Bat Coronaviruses as the Gene Source of Alphacoronavirus and Betacoronavirus and Avian Coronaviruses as the Gene Source of Gammacoronavirusand Deltacoronavirus" Journal of Virology 2011, 3995– 4008; doi:10.1128/JVI.06540-11

[38] P. C. Y. Woo, Y. Huang, S. K. P. Lau and K.Y. Yuen "Coronavirus Genomics and Bioinformatics Analysis"Viruses 2010, 2, 1804-1820; doi:10.3390/v2081803

[39] Q. Li, W. Peng, Y. Ou "Prediction and analysis of key protein structures of 2019-nCoV" Future Virol. (2020)15(6), 349–357; doi: 10.2217/fvl-2020-0020

[40] S. F. Altschul, T. L. Madden, A. A. Schaffer et al., "Gapped BLAST and PSI-BLAST: a new generation of protein database search programs," Nucleic acids research, vol. 25, no. 17, pp. 3389–3402, 1997

[41] S. Kumar, G. Stecher, M. Li, C. Knyaz, and K. Tamura,"MEGA X: molecular evolutionary genetics analysis across computing platforms," Molecular Biology and Evolution, vol. 35, no. 6, pp. 1547–1549, 2018

[42] S.Dong, J.Sun, Z.Mao, L.Wang, Yi-Lin Lu, Jiesen L "A guideline for homology modeling of the proteins from newly discovered betacoronavirus, 2019 novel coronavirus (2019-nCoV)." J Med Virol. 2020;1–7; DOI:10.1002/jmv.25768

[43] S. Williams, "A Brief History of Human Coronaviruses" June 2, 2020 https://www.the-scientist.com/news- opinion/a-brief-history-of-human-coronaviruses-67600

[44] S.Kanga, M. Yang , Z.Hong , L. Zhang , Z. Huang , X. Chen , S. He , Z. Zhou , Z. Zhou , Q. Chen

, Y.Yan , C.Zhang , H.Shan, S.Chen "Crystal structure of SARS-CoV-2 nucleocapsid protein RNA binding domain reveals potential unique drug targeting sites" Acta Pharmaceutica Sinica B 2020;10(7):1228e1238; https://doi.org/10.1016/j.apsb.2020.04.009

[45] S.Chaplin, "COVID-19: a brief history and treatments in development" prescriber 2020; https://doi.org/10.1002/psb.1843

[46] T.Schwede, J.Kopp, N.Guex and M.C.Peitsch "SWISS-MODEL: an automated protein homology-modeling server" Nucleic Acids Research, 2003, 31: 13 3381–3385 DOI: 10.1093/nar/gkg520

[47] WebMd, "Coronavirus 2020 Outbreak: Latest Updates" https://www.webmd.com/lung/news/20200124/coronavirus-2020-outbreak-latest-updates

[48] WHO https://www.who.int/news-room/detail/07-04-2020-who-lists-two-covid-19-tests-for-emergency- use

[49] Z.Zhu , X.Lian, X.Su , W.Wu, G.A. Marraro and Y. Zeng "From SARS and MERS to COVID-19: a brief summary and comparison of severe acute respiratory infections caused by three highly pathogenic human coronaviruses" Respiratory Research (2020) 21:224 https://doi.org/10.1186/s12931-020-01479-w

[50] Y.Zhanga, X.Gengb, Y.Tana, Q.Lic, C.Xu J.Xu, L.Hao, Z.Zeng, X.Luod, F.Liu, H.Wanga "New understanding of the damage of SARS-CoV-2 infection outside the respiratory system" Biomedicine & Pharmacotherapy (2020); 127.https://doi.org/10.1016/j.biopha.2020.110195

[51] Kanta Subbarao, and Siddhartha Mahanty "Respiratory Virus Infections: Understanding COVID-19" Immunity. (2020); 52(6): 905–909. 10.1016/j.immuni.2020.05.004

IOP Conf. Series: Materials Science and Engineering 1170 (2021) 012004 doi:10.1088/1757-899X/1170/1/012004

Thermal analysis of multi-walled carbon nanotubes doped glass fiber reinforced polymer composites

S K Chaudhary[1] and K K Singh[2]

[1]Department of Mechanical Engineering, B I T Sindri, Dhanbad, India
[2] Department of Mechanical Engineering, IIT(ISM), Dhanbad, India

E-mail: skchaudhary.me@bitsindri.ac.in

Abstract Thermal characterization is one of the important aspects for evaluating the heat flux and decomposition kinetics of different materials, alloys and composites at different temperature. In this work Multi-scale nano-composites (MWCNTs/glass fiber/epoxy) were prepared by hand lay-up process followed by vacuum bagging method. Thermal properties of MWCNTs doped glass fiber reinforced polymer composites were evaluated by differential scanning calorimetry (DSC), thermo gravimetric analysis (TGA) and differential thermal analysis (DTA). It has been found that in the presence of MWCNTs it is very difficult to break the polymeric chains and that leads to reduction in cross linking density and glass transition temperature of MWCNTs doped glass fiber reinforced polymer composites as compared to pristine one.

Keywords: Thermal characterization; Nano-composites; Multiwall carbon nanotubes; Thermo-gravimetric analysis; Differential scanning calorimetry

1. Introduction

Generally composites have superior properties than that of metallic materials. Especially, glass fiber reinforced polymer is widely used by many industries due to its high specific strength, low weight, low cost, high stiffness and corrosion resistance [1, 2]. Moreover, when carbon nanotubes are introduced into composite materials, it will further enhance the overall properties owing to high tensile modulus of carbon nanotubes. Carbon nanotubes may be of either single wall or multiwall in nature. These carbon nanotubes may be produced by arc discharging process, chemical vapour deposition method and laser ablation method. Multi-walled carbon nanotubes are widely used as first phase or second phase reinforcement in polymer composites in order to enhance the mechanical, chemical and electrical properties.

Thermal analysis is one of the important characterization for evaluating decomposition kinetics, thermal stability, enthalpy, heat flux and glass transition temperature of different materials, alloys and composites at different temperature. Thermal stability is very important for any materials, polymer or composites so that it can be used without degradation at certain temperature. Most common thermal analysis such as TGA, DSC and TDA are widely used for characterizing the polymeric materials [3]. Thermo gravimetric analysis (TGA) is used to measure change in weight % with respect to temperature. Degradation of the materials largely depends upon the material structure and heating condition. Thermo -gravimetric analysis is widely used in the field of characterization of different types of polymer nano-composites [4]. Differential scanning calorimetry (DSC) is used to find the melting temperature and glass transition temperature. Glass transition temperature (T_G) is closely related with mobility of polymeric chain which adjudicates the transition phase between glassy and rubbery polymeric state [5].

Present work consists of different thermal analysis (TGA, DSC and DTA) of MWCNTs doped glass fiber reinforced polymer composites. This thermal analysis needs for encompasses the composite materials at a particular temperature without degradation.

2. Experimental

2.1 Materials

Plain weave glass fibers having a weight of 600 GSM were procured from M. S. Industries, Kolkata (India). Density of used glass fiber was 2540 kg/m^3. Bisphenol-A diglycidyl ether (DGEBA) under trade name of Lapox (L-12) was used as a polymer matrix. Triethylene tetra amine based hardener (K6) was used as a hardener. Both materials were procured by Atul industries Ltd, Balasad, Gujrat (India). Multi-walled carbon nanotubes were produced by arc discharge method [6] for used as second phase reinforcement.

2.2 Composite materials manufacturing

Multi-scale nano-composites were manufactured by hand lay-up process and followed by vacuum bagging technique. In hand lay-up process, pre-calculated amount of multi-walled carbon nanotubes were weighed and mixed with matrix material in a beaker by mechanical process. Then, hardener was blended in a beaker. Each laminates (280 x 240 mm^2) of glass fiber was impregnated with mixed materials with matrix and hardener in the mass ratio of 10: 1. Now each impregnated layers was placed one by one and consolidated by hand roller so that entrapped air and extra resin can be removed. In this way total eight plies with different stacking sequence were made. In vacuum bagging method, first of all laminates was kept and seal properly inside the vacuum bag. Then, a pressure of 0.933 bar was applied in order to eliminate voids, entrapped air and extra resin. In the last, specimens were put under load for about 24 hours for complete the curing process.

2.3 Thermo-gravimetric (TGA) and differential thermal analysis (DTA)

Thermal analysis (TGA-DTA) was carried out to check the thermal stability and decomposition kinetics of MWCNT/GFRP samples from room temperature to 800^{0}c. TGA-DTA of samples were recorded on instrument with model no. NETZSCH Jupiter STA 449F3 in the temperature range of 0-800^0 C. A sample size of 15.231 mg was taken and placed in alumina crucible with a heating rate of 10^{0}c/ minute, and heated in atmosphere of nitrogen having flow rate of 20 ml/minute.

2.4 Differential Scanning Calorimetry (DSC)

DSC analysis of different GFRP samples was carried out on instrument DSC Q10 V9.4 Build 287 in order to investigate the effect of temperature rise on polymeric /amorphous materials. Glass transition temperature (T_G) was also evaluated from DSC curve to observe the effect of adding multi-walled carbon nanotubes in glass fiber reinforced polymer laminates. Aluminum pan was used for placing the samples which having weight of 1.556 mg. Heating rate was taken as 10^{0}c/ minute and nitrogen was used as a purge gas.

3. Results and Discussion

3.1 Thermo gravimetric analysis (TGA) for Laminates

TGA curves are used to evaluate the thermal degradation and thermal stability of constituent materials. TGA curves of pristine and MWCNTs doped GFRP laminates are illustrated in 'figure 1'. Thermal degradation of materials largely depends upon the material structure and heating conditions. It can be demonstrated that thermal decomposition of each composite samples was taken place between 30^{0}c to

800^0c.

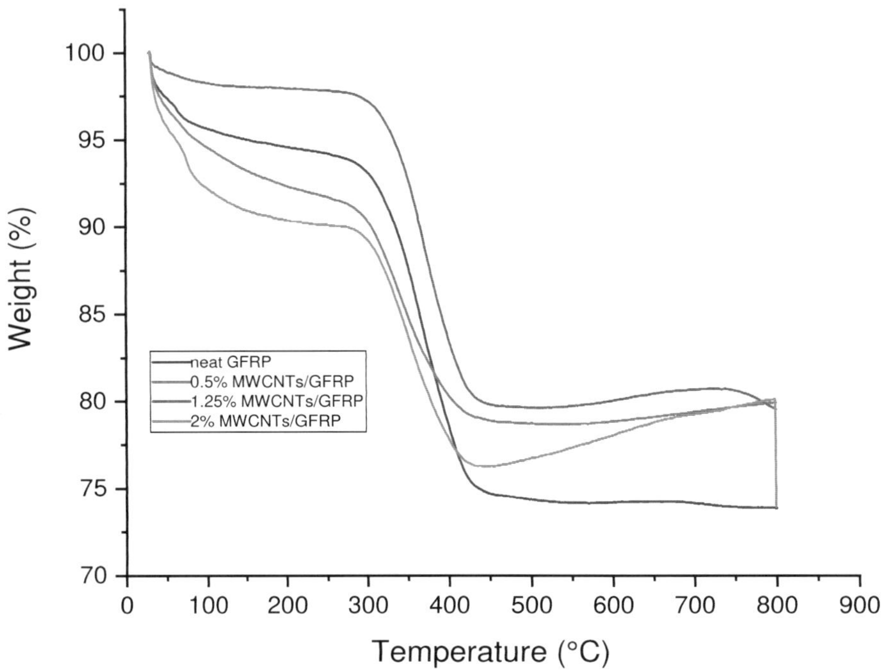

Figure 1. TGA of neat and MWCNTs embedded GFRP laminates

Major degradation occurred almost in the range of 310-430^0c. Substantial and minimum weight losses were found to be 21.2% and13.4% for pristine GFRP and 0.5% MWCNT doped GFRP respectively. 1.25% MWCNTs doped GFRP revealed 18.36% weight loss, while 2% MWCNTs doped GFRP showed 14.44% weight loss. It can be construed that weight loss may be occurred due to thermal degradation of polymer matrix and glass fiber. It can be also delineated that, it is somewhat difficult to break the polymeric chains in the presence of multi-walled carbon nanotubes which leads to reduction in weight loss of MWCNT doped GFRP as compared to pristine GFRP. It can be also derived from TGA curve that between 420-800^0c thermal degradation become stable. Rahman et al. [2] reported that thermal decomposition of neat woven glass fiber reinforced composite started earlier than CNTs grown woven glass fiber reinforced composite. However, Fawad Tariq et al. [7] reported higher weight loss in 0.25 wt% MWCNTs-CFRP as compared to neat carbon fiber reinforced polymer (CFRP) and construed that it may be occurred due to evaporation of residual solvent.

3.2 DTA analysis for laminates

DTA curve of pristine and MWCNTs doped GFRP laminates are illustrated in 'figure 2'. All laminates have approximately coherent nature except 0.5 wt% MWCNTs doped GFRP laminates. However, this results doesn't exhibit any significant peak corresponds to weight loss.

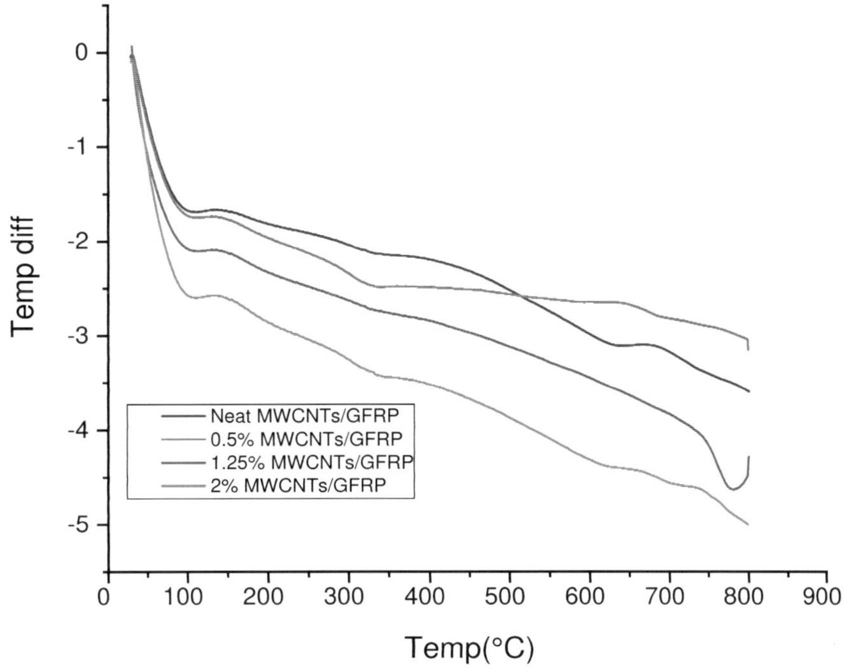

Figure 2. DTA of neat and MWCNTs embedded GFRP laminates

This is of particular interest, but for deeper look into that, it needs further investigation of subsequent derivatives [4].

3.3 DSC analysis for laminates:

DSC curve of MWCNTs doped GFRP is presented in 'figure 3', which clearly demonstrate that pristine GFRP and 2 wt% MWCNT doped GFRP have approximately akin heat flux profile. Coherent heat flux profile can be also seen for 0.5% and 1.25 wt% MWCNT doped GFRP. It can be derived from graph that 0%, 0.5%, 1.25% and 2 wt% MWCNTs doped GFRP have glass transition temperature of 340^0c, 320^0c, 300^0c and 308^0c respectively. This result shows that glass transition temperature of 0.5%, 1.25% and 2 wt% MWCNTs doped GFRP is found to be lower than that of pristine GFRP. It can be also deduce that more reduction in glass transition temperature (T_G) can be found for 1.25 wt% MWCNTs doped GFRP.

Basically, multi-walled carbon nanotubes are interrupted between the polymer molecular chains which restrain to make polymerization and cross linking. Therefore, it can be deduced that, reduction in glass transition temperature (T_G) for MWCNTs doped GFRP may be attributed to reduction in polymerization rate and cross linking density as well as due to pristine nature of multi-walled carbon nanotubes. It can be also observed that neat GFRP exhibited pronounce glass transition temperature as compared to others and this may be attributed to more cross linking formation among polymeric chains in the absence of multi-walled carbon nanotubes. Coherent heat flux profile has been reported by Rahman et al. [2] for neat and CNTs grown woven glass fiber reinforced epoxy composites. Nevertheless, they found some minor change in glass transition temperature (T_G) for CNTs grown woven glass fiber as compared to neat one. However, this results depict 5.88%, 11.76% and 9.41% change in glass transition temperature for 0.5%, 1.25% and 2 wt% MWCNTs doped glass fiber reinforced polymer composite as compared to neat one.

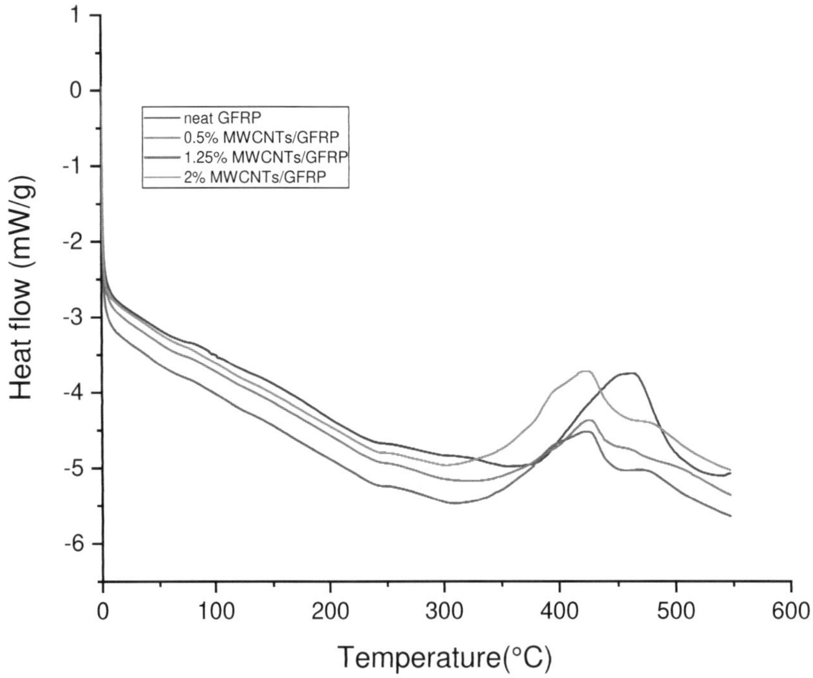

Figure 3. DSC of neat and MWCNTs embedded GFRP laminates

It can be envisaged that multi-walled carbon nanotubes may trap free radicals [8], resulting in reduction of cross linking density of epoxy resin, which actually incapacitate the polymer matrix interface. Rathore et al. [9] also reported reduction in glass transition temperature (T_G) for epoxy/0.1% MWCNT nanocomposite as compared to neat epoxy. It was construed that that MWCNTs may be penetrated into inter chain spacing of polymer that caused restriction in formation of cross link which resulted in reduction in glass transition temperature (T_G).

4. Conclusions

Thermal behaviour of MWCNTs doped GFRP has been successfully studied. It was found that there was a reduction in weight loss of MWCNT doped GFRP as compared to pristine GFRP. This may be due to difficulty in breaking the polymeric chains in the presence of multi-walled carbon nanotubes. DSC analysis reveals that neat GFRP has a higher glass transition temperature than that of MWCNTs doped GFRP composites. This may be ascribed to interruption of MWCNTs between polymeric chains which restrict the mobility of polymeric chains. This causes reduction in cross linking density. DTA analysis doesn't show any significant peak.

Acknowledgements

Author would like to thanks department of Mechanical Engineering, IIT(ISM), Dhanbad and department of Mechanical Engineering, B. I. T. Sindri for providing research facilities.

IOP Conf. Series: Materials Science and Engineering 1170 (2021) 012004 doi:10.1088/1757-899X/1170/1/012004

References

[1] Chaudhary S K, Singh K K and Venugopal R 2018 Experimental and numerical analysis of flexural test of unfilled glass fiber reinforced polymer composite laminate, *Materials Today: Proceedings* **5** 184-192

[2] Rahman S R, Rangari V, Jeelani S 2015 Thermal and Mechanical Properties of Woven Glass Fiber Reinforced Epoxy Composites with Carbon Nanotubes Grown in-Situ, *International Journal of Engineering And Science* **4** 2 54-6

[3] Schindler A, Doedt M, Gezgin S, Menzel J and Schmolzer 2017 Identification of polymers by means of DSC, TG, STA and computer-assisted database search *J Therm Anal Calorim* **129** 833-42.

[4] Ferreira J M, Errajhi O A Z, Richardson M O W 2006 Thermogravimetric analysis of aluminised E- glass fibre reinforced unsaturated polyester composites, *Polymer Testing* **25** 2 1091-94.

[5] Siregar J P, Salit M S, Rahman M Z A, and Dahlan K Z H M 2011 Thermogravimetric Analysis (TGA) and Differential Scanning Calometric (DSC) Analysis of Pineapple Leaf Fibre (PALF) Reinforced High Impact Polystyrene (HIPS) Composites, *Pertanika J. Sci. & Technol.* **19** 161-170

[6] Singh K K, Chaudhary S K, Venugopal R and Gaurav A 2017 Bulk synthesis of multi-walled carbon nanotubes by AC arc discharge method, Proceeding of Institution of Mechanical Engineers, *Part N: Journal of nanomaterials, nanoengineering and nanosystem* **231** 141-151.

[7] Tariq F, Shifa M, Baloch R A 2018 Mechanical and Thermal Properties of Multi-scale carbon Nanotubes- Carbon Fiber-Epoxy Composite, *Arabian Journal for Science and Engineering.* **43** 5937-48.

[8] Zhu J, Imam A, Crane R, Lozano K, Khabashesku and Barrera E V 2007 Processing a glass fiber reinforced vinyl ester composite with nanotube enhancement of interlaminar shear strength, *Compos Sci Technol.* **67** 1509-17

[9] Rathore D K, Prusty R K, Kumar D S and Ray B C 2016 Mechanical performance of CNT-filled glass fiber/epoxy composite in in-situ elevated temperature environments emphasizing the role of CNT content, *Composites: Part A* **84** 364-376.

IOP Conf. Series: Materials Science and Engineering 1170 (2021) 012005 doi:10.1088/1757-899X/1170/1/012005

Design of Planetary Gear Train for Geared Rotary Actuator

Yadnik Kude[1, a)] Priya Gajjal[2, b)] Anuj Kumar Singh[3, c)]

[1]*Yadnik V Kude, Mechanical Engineering, AISSMS COE, Maharashtra, India*
[2]*Priya S Gajjal, Mechanical Engineering, AISSMS COE, Maharashtra, India*
[3]*Anuj Kumar Singh, R&D Engineers, Maharashtra, India*

.

[a)] *yadnikude@gmail.com*
[b)] *singhanuj81@gmail.com*
[c)] *psgajjal@aissmscoe.in*

Abstract: Geared Rotary Actuators are mechanically operable devices widely used in the aerospace industry so that controlled motion can be provided to secondary flight control surfaces. Their usage can be seen in a variety of applications including powered hinges for aircraft structural movements, they support aerodynamic loads in addition to the surface hinge moment, as an actuator in a linked mechanism where they supply actuation torque. Mostly GRAs are situated along with the bay door drives, and also can be used for controlling leading or trailing edge movements of the aircraft. They receive Power Drive Unit (PDU) torque via assembly of shafts and joints. From a mechanical standpoint, the GRAs can be termed as epicyclical gear reducer or compound planetary gearbox. In this work, the governing criteria and mathematical equations used to characterize GRA are discussed. SOLIDWORKS is used for part modeling of GRA. More focus will be on the speed reduction ratio calculation, determination of strength, and overall Factor of Safety (FOS) of the torque transmission unit. Thus an attempt is made to obtain the best performance parameters to meet the desired requirements of Geared Rotary Actuator.

Keywords: *Geared Rotary Actuators, secondary flight control surfaces, powered hinge, power drive unit, SOLIDWORKS, factor of safety*

1. INTRODUCTION

An Actuator is a part of a machine that is mainly responsible for moving and controlling a mechanism or a complete system. An actuator receives input as a control signal and a source of energy. A control signal is relatively low energy and maybe electric voltage or current, pneumatic or hydraulic pressure, or can be a human-generated power. When it receives a control signal, an actuator reacts by converting the signal's energy into desired motion.

Mechanical Actuators are generally used to translate mechanical motion (often rotary) into linear motion or vice versa with the help of gearing arrangements. These actuators are typically part of a larger system which can be power drive units, mechanical interconnects, and feedback devices to control the motion of multiple devices.

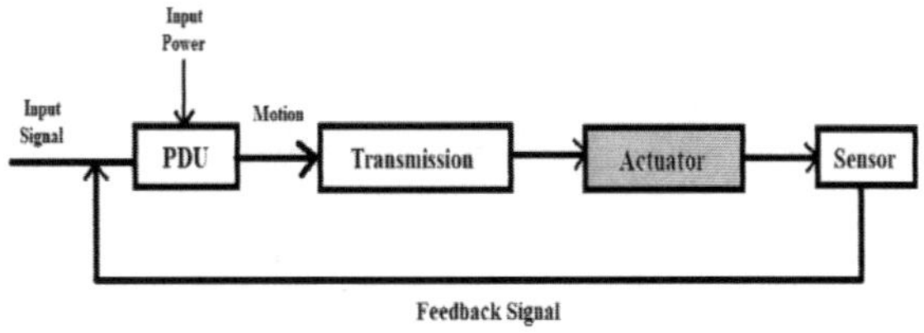

FIGURE 1. Block Diagram of Transmission System

1.1 Geared Rotary Actuator (GRA)

Geared Mechanical Actuator is a geared mechanism that is used to convert mechanical motion (often rotary) into rotary motion at a different combination of speeds and force. They are also used when required to change the direction of motion. Unlike hydraulic or electromechanical actuators, they receive their power from an external source. Further GRA is classified as simple planetary actuators and compound differential planetary actuators. Simple planetary actuators are most commonly found in commercial leading-edge slat applications and are bolted to structure driving rack and pinions to translate the slat surface. Compound differential planetary actuators are most commonly used in trailing edge flap designs. It offers higher ratios for torque multiplication while driving rack and pinions to translate the flap surfaces.

Geared Rotary Actuators are initially used for the positional control of wing-mounted flight control Surfaces of an aircraft. Their usage ranges from their application in powered hinges, when directly supporting the aerodynamic load forces in addition to the surface hinge moment, to actuators in a linked mechanism where they provide the actuation torque and the aerodynamic load reaction and is delegated to other load-bearing equipment. In almost all cases, the GRAs are situated along with the leading or trailing edges of an aircraft and are connected to each other via shafts and joints while transferring the Power Drive Unit torque to the GRAs.

1.2 Working of GRA

Geared Rotary Actuators (GRA's) can be used to control the movement of leading-edge flaps of aircraft wing Structures to alter the lift characteristics of the wing. Each actuator includes an epicyclic-type reduction gearing arrangement which transmits the rotation of an input member to a pair of output members. One of which is fixed to the wing Structure, and So is usually referred to as an "earth' and another one is coupled to the control Surface to move the control Surface relative to the wing Structure of an aircraft.
Geared rotary actuators (GRA) are mechanical assemblies performing speed reduction and torque amplification from the input shaft to the actuator output and providing reaction to the control surface loads. The compound planetary gear arrangement differs from a conventional epicyclic gear arrangement by eliminating the carrier assembly for planetary gears. This is made possible by balancing the tangential teeth forces on the planetary pinions at the ring gear meshes.

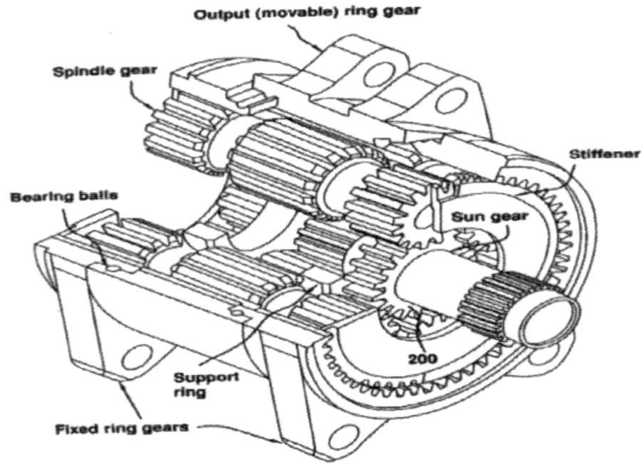

FIGURE 2. Model of GRA [2]

The compound planetary gear set contains a sun gear receiving the input torque through the input shaft, a ring gear supplying the output torque to the driven element, two stationary ring gears symmetrically placed at the two sides of the output ring gear. The power flows from the input sun gear to the output ring gear through multiple planet gears meshing around the complete periphery of the sun gear. Each planet gear consists of three pinions: the outer end pinion gears are identical to each other and are meshing with two fixed outer ring gears. For each planet gear, the tooth load at the center mesh is reacted by the two identical forces at the fixed meshes. A full arrangement of planetary gears is used to distribute the load over a large number of gear teeth which, in addition to minimizes the stresses on the gear tooth and ensures more even load distribution which ultimately helps to increase the stiffness of an actuator.

The output element of the geared rotary actuator is radially supported by the fixed housing. Radial support occurs along two parallel circular surfaces, symmetrically located at the two sides of the actuator output. Such type of support provides the necessary radial reaction force to balance the force acting on the actuator output. While the moment of the external force about the actuator axis is balanced by the torque provided by the actuator output gear. As a result, the actuator behaves as a powered hinge.

The Power Hinge can be subjected to torque, shear, and axial thrust loads. The hinge shear and thrust loads are isolated from the hinge moment (torque) by the annular ball bearings coupling the movable output ring gear to the fixed ring gears. Since the fixed ring gears are typically supported by aircraft structure, all loads originating at the output ring gear are transmitted through the annular ball bearings to the fixed ring gear and the aircraft structure. Thus, the spindle gears do not experience unbalanced loads; only torque about the axis of the actuator.

2. NUMERICAL MODEL OF GRA

As for epicyclic gear reducers, gears rotating about a fixed axis when observed from a static observer are called sun gears, while gears whose axis is not stationary are defined planets underlining their revolving motion around the sun gears. The gears making up the rotary actuators are spur gears which are sized according to the module and the number of teeth. An important feature of the teeth commonly used in the GRA planets is the rounded axial profile. This non-standard geometry accommodates the planets deformation due to the bending moment, thereby ensuring a more uniform load distribution along the teeth length that contributes to reduce localized stress. It is also important to underline that most often the gears of a GRA are cut with a large correction increasing the root thickness permitting higher load transfer in gears with a limited number of teeth (whole assembly must be compact as possible and should be light in weight). Desired Technical Specifications of this GRA are as follows:

- Torque Required at the output (to be provided by moving ring gear) = 4500 N-m at 20 RPM
- Speed available at input of sun gear = 2964 RPM
- Desired Reduction Ratio = 148.2:1
- Desired Time of Operation = less than 2 secs

2.1 Calculations for Reduction Ratio

The speed ratio between output and input of a GRA can be calculated by applying repeatedly the well-known Willis equation used for ordinary epicyclic gear reducers.

$$r = \frac{\omega_n - \omega_p}{\omega_1 - \omega_p}$$

$$r = \frac{\frac{Z_2 Z_5}{Z_3 Z_4} - 1}{\frac{Z_2 Z_5}{Z_1 Z_3} + 1} \quad \text{.........Willi's equation}$$

As we are designing this GRA for reduction ratio purpose, it can be obtained by taking inverse of Gear Ratio as,

$$\text{Reduction Ratio} = \frac{1}{\text{Gear Ratio}} = \frac{1}{r}$$

For obtaining correct value of reduction ratio we must know the values of number of teeth on sun, planetary pinion and ring gear respectively. This can be obtained by fulfilling the following conditions while selecting number of teeth. There are three essential conditions for the planetary gear train to transmit the power from input shaft to output shaft. These essential conditions are as:

- **Condition-1:**
  This expression is valid for the gear system designed as standard gear system. No. of teeth on the gear can be modified by gear design based on profile shift. Centre distance between two gears to be matched. i.e. $a_1 = a_2$
  $$Z_4 = Z_1 + 2Z_2$$
- **Condition-2:**
  It is requisite for placing the planet gears consistently about the sun gear.
  $$\frac{Z_1 + Z_4}{N} = I$$
- **Condition-3:**
  It ensures the operation of the planet gears without any interference with each other. This condition is the requisite for meeting the standard gear design where planet gears are equally placed.
  $$Z_2 + 2 < (Z_1 + Z_2) \sin\left(\frac{180}{N}\right)$$
- **Condition-4:**
  In such compact designs while operating at such higher reduction ratios, gears in mesh may subject to higher wear and undercutting problems. So, this condition ensures that while operating, gears would not undercut.
  $$(Z_3 + 2) < \left(\frac{z_1' + z_3}{2}\right)$$

These four conditions are necessary for any planetary gear train to transmit power and motion. Any of these conditions not getting satisfied will result in non-feasible planetary gear train design.
After satisfying all the above conditions and performing satisfying number of iterations we finalised number of teeth values as follows,

- Number of teeth on Sun Gear, Z_1 = 35
- Number of teeth on Sun Gear Spline, Z_1' = 34
- Number of teeth on Central Gear of planetary pinion, Z_2 = 28
- Number of teeth on Outer Gear of planetary pinion, Z_3 = 29
- Number of teeth on Moving Ring Gear, Z_4 = 91

Number of teeth on Fixed Ring Gear, Z_5 = 92

This combination of teeth satisfies all the above four mentioned conditions, as well as it gives desired high reduction ratio without any undercutting of gears.

Condition-1: **Satisfied**

$$Z_4 = Z_1 + 2Z_2$$
$$91 = 35 + (2*28)$$
$$\mathbf{91 = 91}$$

Condition-2: **Satisfied**

$$\frac{Z_1 + Z_4}{N} = I$$
$$\frac{35 + 91}{6} = 21$$
$$\mathbf{21 = 21}$$

Condition-3: **Satisfied**

$$Z_2 + 2 < (Z_1 + Z_2)\sin\left(\frac{180}{N}\right)$$
$$\mathbf{30 < 31.5}$$

Condition-4: **Satisfied**

$$(Z_3 + 2) < \left(\frac{z_1' + z_3}{2}\right)$$
$$(29 + 2) < \left(\frac{34 + 29}{2}\right)$$
$$\mathbf{31 < 31.5}$$

Now, reduction ratio is given by

$$r = \frac{\frac{Z_2 Z_5}{Z_3 Z_4} - 1}{\frac{Z_2 Z_5}{Z_1 Z_3} + 1}$$

$$r = \frac{\left(\frac{28 * 92}{29 * 91}\right) - 1}{\left(\frac{28 * 92}{35 * 29}\right) + 1} = -0.006747638$$

So, **reduction ratio**$= \frac{1}{r} = \frac{1}{-0.006747638} \approx \mathbf{-148.2:1}$

Here we achieved the desired reduction ratio and with this we are able to provide 4500 N-m torque as an output via moving ring gear. As we know that, this whole compound planetary assembly is used for converting high-speed low-torque input into low-speed high-torque output with one of its ring gear connected to flight control surfaces while other remain fixed and bolted to the aircraft structure. So during operation, whole planetary gear assembly might be subjected to various stresses and hence, it should have

sufficient strength so that torque will be transmitted efficiently without failure of any parts as well as it should possess sufficient value of Factor of Safety.
From the observations it has been seen that spur gears are most likely subjected to bending and wear type of failures. Out of the two, industrial gears are mostly designed with the consideration of wear failure only. Hence, we designed gears with the consideration of wear strength criteria while determining desired value of Factor of Safety (FOS).

2.2 Material used, Dimensions of Gears and FOS Calculations

- **Material to be used** = AISI 440C / ASTM S44004 / X105CrMo17
  Application Areas = Antifriction Bearings and Races, Structural Uses in Corrosive environments, Heavy duty industrial gears, Transmission Shafts, Aircraft Forgings
  Ultimate tensile strength = 2170 MPa
  Yield Tensile Strength = 2090 MPa
  Type of heat Treatment = Martensitic Stainless steel - Tempered at 316 ^{0}C

- **Dimensions of Gears**

Module (m) = 1.75
Pressure angle (α) = 20^0
Centre distance,

For external gears, (For 1-2 pair)

$$\text{C.D1}=\left(\frac{Z_1+Z_2}{2}\right)*m=\frac{35+28}{2}*1.75=55.125\text{ mm}$$

For internal gears, (For 2-4 and 3-5 pairs)

$$\text{C.D2}=\left(\frac{Z_4-Z_2}{2}\right)*m=\frac{91-28}{2}*1.75=55.125\text{ mm}$$

Addendum, $h_a = 1*m = 1.75$
Tooth depth (h) = 2.25*m = 3.9375
Minimum Rim Thickness,

$\mathbf{t_r(min)}$ = 1.2*(Tooth Depth) = 1.2*3.9375 = 4.725 mm
So, minimum rim thickness while designing the gears should be at least 4.725 mm.

Pitch Circle Diameter ($\mathbf{D_p}$)
$D_{p1} = m * Z_1 = 61.25$ mm
$D_{p2} = m * Z_2 = 49$ mm
$D_{p3} = m * Z_3 = 50.75$ mm
$D_{p4} = m * Z_4 = 159.25$ mm
$D_{p5} = m * Z_5 = 161$ mm

Tip Diameter ($\mathbf{D_a}$)
$D_{a1} = D_{p1} + (2*m)$ =64.75 mm
$D_{a2} = D_{p2} + (2*m)$ =52.5 mm
$D_{a3} = D_{p3} + (2*m)$ =54.25 mm
$D_{a4} = D_{p4} - (2*m)$ =155.75 mm
$D_{a5} = D_{p5} - (2*m)$ =157.5 mm

Root Diameter ($\mathbf{D_f}$)
$D_{f1} = D_{a1} - (2*h)$ =56.875 mm
$D_{f2} = D_{a2} - (2*h)$ =44.625 mm
$D_{f3} = D_{a3} - (2*h)$ =46.375 mm
$D_{f4} = D_{a4} + (2*h)$ =163.625 mm

$D_{f5} = D_{a5} + (2 * h) = 165.375$ mm

- **Factor of Safety (FOS)**

FOS= $\frac{S_b}{P_{eff}}$

FOS= $\frac{S_w}{P_{eff}}$

Effective Load acting on Gear Tooth (Peff):

$$\mathbf{P_{eff} = \frac{C_s}{C_v} * P_t}$$

Pitch Line Velocity (v)

$= \frac{\pi DN}{60000} = \frac{\pi * 52.5 * 2964}{60000} = 9.505674\ m/s$

Velocity factor (C_v)

$= \frac{3}{3+v} = 0.2398911$ (as $v < 10m/s$)

By referring table no.1 value of service factor ($C_{s)}$ can be taken as, 1.75.

TABLE 1. Service Factor for speed reduction Gearboxes

Working Characteristics of Driving Machine	Working Characteristics of Driven Machine		
	Uniform	Medium Shock	Heavy Shock
Uniform	1	1.25	1.75
Light Shock	1.25	1.5	2
Medium Shock	1.5	1.75	2.25

Now, tangential force due to rated torque can be found as: (Considering 6 planet gears)

$$P_t = \frac{2 * M_t}{6D_{p1}} = \frac{2 * 30364.37}{6 * 61.25} = 165.2482721\ N$$

Effective load acting on gear tooth can be given as,

$$\mathbf{P_{eff}} = \frac{1.75}{0.239891} * 165.2482721 = \mathbf{1205.48226\ N}$$

Beam Strength of Gear tooth (Sb):
(For 1 and 2 gear pair)

$$\mathbf{S_b} = m * b * Y * \sigma_b$$
$$= 1.75 * 35 * 0.373 * 723.33$$
$$= \mathbf{16525.45417\ N}$$

Factor of Safety for 1-2 pair is,

$$\mathbf{FOS} = \frac{S_b}{P_{eff}} = \frac{16525.45417}{1205.48226} = 13.709$$

Wear strength of gear tooth (S_w):

(For 1 and 2 gear pair)

$$S_w = D_{p1} * b * Q * k$$
$$S_w = D_{p1} * b * Q * 0.16 * \left(\frac{BHN}{100}\right)^2$$

Where, ratio factor (Q) = $\frac{2*Z_g}{Z_g+Z_p} = \frac{2*35}{35+28} = 1.111$

$$\mathbf{S_w} = 61.25 * 35 * 1.11 * 5.9536 = \mathbf{14181.144\ N}$$

Factor of Safety for 1-2 pair is,

$$\mathbf{FOS} = \frac{S_w}{P_{eff}} = \frac{14181.144}{1205.48226} = 11.764$$

But, in this case factor of safety is very high. Hence, as per the standards for industrial purpose, here we have used Buckingham's equation for attaining required factor of safety (FOS).

Buckingham's equation for dynamic load (P_d)

$$P_d = \frac{21v(c * e * b + P_t)}{21v + \sqrt{(c * e * b) + P_t}}$$

Where, c is deformation factor and e is sum of errors between two Meshing Teeth can be determined by using table no. 2 and table no.3 respectively.

TABLE 2. Values of Deformation Factor

Materials		14.5° full depth Teeth	20° full depth Teeth	20° Stub Teeth
Pinion Mtl.	**Gear Mtl.**			
Grey CI	Grey CI	5500	5700	5900
Steel	Grey CI	7600	7900	8100
Steel	Steel	11000	11400	11900

TABLE 3. Tolerances on the Adjacent Pitch

Grade	e (microns)
1	0.80 + 0.06φ
2	1.25 + 0.10φ
3	2.00 + 0.16φ
4	3.20 + 0.25φ
5	5.00 + 0.40φ
6	8.00 + 0.63φ
7	11.00 + 0.90φ
8	16.00 + 1.25φ
9	22.00 + 1.80φ
10	32.00 + 2.50φ
11	45.00 + 3.55φ
12	63.00 + 5.00φ

To determine sum of errors between two meshing teeth (e) we used Grade 3 tolerance on adjacent pitch, which will be used further to determine dynamic load (P_d).

$$e = 2.00 + 0.16\varphi$$

Now, Total error (e) is given by,

$$e = e_p + e_g$$

Where,

$$e_p = \text{Error for Pinion}$$
$$e_g = \text{Error for Gear}$$

Where φ is Tolerance Factor and generally taken as,

$$\varphi = m + 0.25\ \text{SQRT}\ (D_p)$$

$$\mathbf{P_d} = \frac{21 * 9.5056(11400 * 0.0052 * 35 + 165.248272)}{21 * 9.5056 + \sqrt{(11400 * 0.0052 * 35)} + 165.248272}$$

$$= \mathbf{1797.0296\ N}$$

For Buckingham's Equation Effective load can be determined as,

$$P_{eff} = (C_s * P_t + P_d)$$
$$= (1.75 * 165.248272) + 1797.0296$$
$$\mathbf{P_{eff} = 2086.2\ N}$$

In industrial and heavy duty applications gears are mostly subjected to fatigue and repetitive type of stresses. Hence, for such applications spur gears are mostly designed with consideration of wear failure criterion. Hence, FOS can be found as:

$$\mathbf{FOS} = \frac{S_w}{P_{eff}} = \frac{4673.48224}{4535.542} = \mathbf{6.7975}$$

So, here we got comparatively better values of factor of safety. But still there is scope of improvement by changing parameters like face width of gears (b) up to some extent or by changing material used for manufacturing of GRA.

3. RESULTS

As per the given data we need to design a device which will be capable of transmitting very high torque at very low speeds. As per the given data, device must be capable of providing 4500 N-m torque as output with output gear rotating at the speed of 20 RPM. So, here we have designed carrier-less compound Planetary Gear Train which will be capable of providing high torque as an output. While designing gear set which will meet our requirements, we have performed number of iterations to find out suitable number of teeth values so that design will be compact and operable according to our requirements.

a) Using Willi's formula for gear ratio we determined satisfactory value of reduction ratio. i.e. Reduction ratio = -148.2:1
b) Using this reduction ratio value and after satisfying all the tooth combination criteria's obtained values of tooth combinations are as:
   $Z_1=35$, $Z_1'=34$, $Z_2=28$, $Z_3=29$, $Z_4=91$, $Z_5=92$
c) While satisfying all the dimensional criteria's of PGT, it's also required that torque must be transmitted efficiently without any failure of the component as well as without affecting the performance of device. Hence, factor of safety is determined based on wear strength failure criteria for each transmission pair of gears using Buckingham's dynamic load equation. It's as follows:

- For 1-2 gear pair, determined value of factor of safety(FOS) is 6.7975
- For 2-4 gear pair, determined value of factor of safety(FOS) is 2.5292

- For 3-5 gear pair, determined value of factor of safety(FOS) is 2.0742

4. CONCLUSION

One of the prime objectives of this work is to develop a system which is highly compact and light in weight. From the analytical and numerical results we are able to obtain high reduction ratio which made this system possible to provide low-speed high-torque as output. As well as, we obtained satisfying value of Factor of safety while taking into consideration the criteria of wear failure.
From the study it has been observed that Factor of Safety (FOS) can be further improved by altering various parameters. Basic parameters which also affect the FOS are Ultimate strength (U_{ts}) well as yield strength (Y_{ts}) of material, face width of gear, module of gear and service factor of gear. On improving the value of U_{ts}, Y_{ts} and face width of gear, the value of Factor of Safety also starts increasing. Factor of Safety can also be improved by decreasing the value of service factor. But there are some drawbacks of altering these parameters. While improving FOS, one can increase the value face width of gear up to certain extent only. Because while optimizing the value of FOS by changing these parameters, also increases size and ultimately weight of the gear unit. Similarly, on decreasing the value of service factor improves the value of FOS. But on the other hand it also affects the life of Product. Hence, one needs to select the parameters according to desired requirements of the product in such a way that it would not affect other parameters of the design.

5. ACKNOWLEDGEMENT

It gives me really great pleasure to present this paper on topic of "Design of Planetary Gear train for Geared rotary Actuator" here on such a great platform.
I take this opportunity to convey my deepest gratitude to Dr. P. S. Gajjal and Mr. Anuj Kumar Singh. They guided me at every instant of need and without their inspiration and guidance this would have been a very difficult task. I am also thankful to all the PG faculties for their help.

REFERENCES

[1]. Alessandro Bertucci, Giovanni Jacazio and Massimo Sorli, "Performance Study and Mathematical Model of Aerospace Geared Rotary Actuators" *International Journal of Applied Engineering Research*, ISSN 0973-4562 Volume 13, Number 1 (2018) pp. 167-174.

[2]. Stephen John Buxton, "Geared Rotary Actuators", *United States Patent Application Publication,* Pub. No.: US007201700B2, 2007

[3]. Dr. Alexander Kapelevich, "High Gear Ratio Epicyclic Drives Analysis", *GearTechnology*- June 2014

[4]. IS 4460 (parts 1 to 3): 1995, Indian Standard dimensions for involute sided splines

[5]. IS 3665-1996, Indian Standard of Gears, for spur and helical gears

[6]. Robert K. Tiedeman, Wayne; Arnold G. Seipel, Montville; William C. Leuze, Randolph, all of N.J. "Rotary Hinge Actuator" *Curtiss-Wright Flight Systems, Inc., Fairfield, N.J,* Patent number: 4932613, 1990

[7]. Bernd-Robert Höhn, Karsten Stahl and Philipp Gwinner, "Light-Weight Design for Planetary Gear Transmissions", *GearTechnology*- September 2013

[8]. Ulrich Kissling, Inho Bae, "Optimization Procedure for Complete Planetary Gearboxes with Torque, Weight, Costs and Dimensional Restrictions", *Applied Mechanics and Materials Vol. 86 pp 51-54,*2011

[9]. S.B.Nandeppagoudar, S.N.Shaikh, S.H.Gawande, "Optimization of Planetary Gear Box for High reduction ratio", *IOSR Journal of Mechanical and Civil Engineering, e-ISSN: 2278-1684,* March 2017

[10]. Anngwo Wang, Seth E. Gitnes, Dr. Lotfi El-Bayoumy, Jonathan P. Davies "Design failure Analysis of high lift system of an aircraft including trailing or leading edge slats/flaps " *Gear Technology*, August-2014

[11]. "Geometry factors for determining the pitting resistance and bending strength of spur, helical and herringbone gear teeth", AGMA 908-B89, April-1989

[12]. G. M. Maitra Handbook of gear design, pp 2.145-2.147

[13]. V. B. Bhandari, "Design of Machine Elements", pp 646-690

[14]. R. G. Budynas, J. K. Nisbett, "Mechanical Engineering Design" *Shigley's*, pp 665-767

Experimental studies on utilization of Coarser Fly Ash in Concrete

Dr. V. G. Meshram[1], Dr. A.M. Pande[2,b] Dr. S. B. Borghate[3,c]

[1]*Associate Professor, Department of Civil Engg., Yeshwantrao Chavan College of Engineering, Nagpur, India,*
[2]*Professor, Department of Civil Engg., Yeshwantrao Chavan College of Engineering, Nagpur, India,*
[3]*Associate Professor, Dept. of Applied Mechanics, Visvesarayya National Institute of Technology, Nagpur, India;*

[a]*meshramvg@gmail.com,*
[b]*anantpande.ycce@gmail.com,*
[c]*sbborghate@apm.vnit.ac.in*

Abstract: *Fly ash is abundantly available in India. IS 3812 Part I & II regulate its usage in cement and concrete. It is well known fact that fineness of fly ash goes on increasing from boiler end to chimney, however the output goes on reducing. Past research normally indicate usage of finer (Hopper no. 4 & 5) fly ash in cement and concrete. Japan has upgraded its standards and specifications to accommodate all grades of fly ash available. The present research by the authors on all grades of fly ash samples through compressive strength development indicate all grades of fly ash can be used in concrete with a precaution that their performance should be investigated before use. Through graphical representations of test results presented in the paper, it is justified that coarser fly ash can very well be used in concrete making as per strength requirements.*

Keywords: *Coarse fly ash, Fineness, Physical and Chemical Properties, Compressive strength, Regression Analysis*

1. INTRODUCTION

Fly ash is a byproduct which results from the burning of bituminous or sub-bituminous coal. Out of the total fly ash produced, about 80% is finely divided. This fine fly ash gets out to boiler along with flue gases. The Fly ash which is separated out of the flue gases is entrapped in electrostatic precipitators (ESP). The ESP normally consists of four to six fields (hoppers). The field at the boiler end is first and the one at the stack (chimney) end is last hopper. The output of fly ash vary hopper wise which depends on the partial pressures in each hopper of the ESP. The approximate output and fineness of fly ash from hopper from the boiler end to stack end of ESP is explained in Table 1.[1] . This trend is common for all power plants.

Table 1. Hopper-wise output and fineness of Flyash [1]

Field No.(Hopper)	Specific surface area m^2/kg	Approximate output %
1(Boiler end)	229	55-65
2	320	20-25
3	474	5-10
4	576	2-3
5 (Stack end)	607	1-2

Normally fly ash used in concrete mixes is from either hopper No. 4 or 5 and fly ash available in hopper No. 1, 2 & 3 is overlooked in making concrete due to fineness which governs most of the properties as reported by various researchers.

Kalgal M et. al. [2] explained the significance of physical characteristics like particle size distribution and fineness on the reactivity of the ash. The pozzolanic activity of fine fly ash is more as more reactive surface area is available to react with lime [1]. The water demand using fine is reduced as fly ash particles fill the void replacing water within the concrete mix [3,4]. The workability of concrete at a given water cementitious material ratio is improved as smooth surface and the ball bearing effect due to spheroid shape of fly ash accelerate the easy mobility of water [5]. The fine fly ash produces mortars with good strength and low porosity [6]. The inclusion of fly ash increases the workability and air content of concrete while decreasing unit weight and compressive strength of concrete [7]. All these improved properties of concrete lead to the increase in compressive and tensile strength as fly ash fineness increased [8].

The Indian fly ash from hoppers 1, 2 & 3 do not qualify for consideration as an ingredient in concrete mainly due to the stringent provision of IS 3812-2000 Part I & II [15,16]. However, the era of High Volume Fly Ash Concrete supplements the usage of coarser ashes in concrete. Use of all grades of fly ash in concrete is recommended in Japan and The Japanese Code JIS 6201 A supplements the uses. Performance of all grades of fly ash has been recorded in literature from this country (S. Nagataki et al.[9]). V. Sivasundaram [10] showed that almost all of the Indian fly ashes are suitable for use in the high-volume fly ash (HVFA) concrete system. B. Fournier et. al.[11] used coarse fly ash to evaluate their suitability for use in the manufacture of HVFAC.

A.M. Pande and L.M. Gupta [12] discussed the significance of properties of concrete ingredient in the compressive strength development. Chai et. al. [13] compared the compressive strengths and the strength activity indices of fly ash cement mortars and showed that coarse fly ash resulted in slow strength development. V. Sivasundaram [10] showed that the concrete made with the cement and fly ash showed a lower early strength but the strength gain with age was very rapid. Rafat Siddique [14] indicated that the use of high volumes of Class F fly ash as a partial replacement of cement in concrete decreased its 28-day compressive. However, strength properties showed continuous and significant improvement at the ages of 91 and 365 days.

The present paper gives focus on the performance of fly ash from first three hoppers and their comparison with the performance of fly ash from last two hoppers. The outcome of this paper may induce an idea of starting use of the coarse ashes for fly ash concrete.

2. Experimental and Research Methodology

A. Collection: The fly ash under study is collected From Khaparkheda Thermal Power Station, near to Nagpur, Maharashtra. The fly ash was collected from all the hoppers.

B. Physical and Chemical Properties of Fly ash: All samples were tested for physical and chemical properties. For each test minimum three samples were tested. Tests were performed as per provisions of IS 1727[17]. The physical properties of Fly ash are given in Table 2 and the chemical properties of Fly ash are given in Table 3.

Table 2. Physical Properties of Fly ash

Sr. No.	Test conducted	Requirement as per IS: 3812 Part 1-2003	Test Result				
			Hopper No. 1	Hopper No. 2	Hopper No. 3	Hopper No. 4	Hopper No. 5
1	Consistency %	***	27.5	27.5	25	24	25
2	Specific Gravity	***	2.04	2.06	2.17	2.19	2.28
3	Setting Time (In minutes)						
	Initial	***	250	245	240	205	180
	Final		330	325	325	280	270
4	Soundness(mm) By Autoclave Expansion method	Max 0.8	-0.0516	-0.0502	-0.0312	-0.0473	-0.0576
5	Fineness % by weight By Sieving (% Retention on 45 micron sieve - wet sieving)	Not more than 34	45.55	24.45	2.7	6.7	0.5
6	Fineness (Specific Surface) (Sq. m. / kg) By Blain Air Permeability	Min 320	229	320	474	536	607
7	Flow %	***	15.00	14.00	12.00	12.00	12.00
8	Compressive Strength(N/sq. mm)	Not less than 80 % of cement at 28 Days (Min 33)					
	i) 7 Days		19	20	31.5	24	28
	ii) 28 Days		31	34	47	45	57
	iii) 90 Days		46	58	67	67	72

Table 3. Chemical Properties of Fly ash

Sr. No.	Test conducted	Requirement as per IS: 3812 Part 1-2003	Test Result				
			Hopper No. 1	Hopper No. 2	Hopper No. 3	Hopper No. 4	Hopper No. 5
1	Loss on Ignition %	Max. 5.0	1.43	2.52	4.11	1.9	2.09
2	Silicon dioxide (SiO2) in percent by mass	Min. 35	56.02	56.05	53.11	54.63	51.73
3	Silicon dioxide (SiO2) + Aluminium Oxide (Al2O3) + iron oxide (Fe2O3) in percent by mass	SiO2 + Fe2O3 + Al2O3 Min. 70%	88.89	88.76	85.2	86.21	91.21

4	Magnesium oxide (MgO) content- %	Max. 5.0	0.55	0.19	0.54	0.43	0.15
5	Total Sulphur as Sulphur trioxide(SO3) in percent by mass	Max. 3.0	Below 0.1	0.32	0.24	0.7	0.66
6	Calcium oxide(CaO) content-%	***	1.53	0.95	0.29	0.25	1.92

C. Mix Design: Mix design for a concrete was done for various grades as per IS 10262- 2019 [18]. Three grades of concrete namely The M20, M25 and M30 and Grade 43 (OPC) cement was used. The fine and coarse aggregates were used in the Mix Design as per IS 383-2016 [19]. Concrete samples with 0%, 12.5 %, 25 % and 37.5 % replacement of cement with fly ash from each of the hopper 1 to 5 were casted. Water-cement ratio for all percentages of fly ash was kept constant.

D. Compressive strength Test: Testing for compressive strength was carried out in three stages i. e. after 7, 28 and 90 days. The test procedure was as per IS 516-1959(Reaffirmed 2018) [20]. The test results are presented in Table 4.

Table 4. Compressive strength of concrete

Mix Grade	FA %	Hopper No.	Avg. Compressive Strength		
			7 days	28 Days	90 Days
M20	0		20.15	27.71	28.3
M25	0		21.63	35.85	36.89
M30	0		31.11	41.78	43.26
M20	12.50%	1	18.22	26.89	26.23
M20	12.50%	2	17.77	24.6	27.56
M20	12.50%	3	20	24.6	28.15
M20	12.50%	4	19.26	28	29.78
M20	12.50%	5	19.55	28.08	30.37
M20	25%	1	17.92	24.67	28.89
M20	25%	2	18.37	27.27	29.78
M20	25%	3	19.11	28.3	31.56
M20	25%	4	19.26	26.82	31.11
M20	25%	5	20	27.71	31.85
M20	37.50%	1	17.92	22.97	25.34
M20	37.50%	2	17.77	25.34	26.52
M20	37.50%	3	18.07	24.6	26.97
M20	37.50%	4	20.11	24.3	28.44
M20	37.50%	5	18.22	26.82	29.19
M25	12.50%	1	17.63	30.82	34.79
M25	12.50%	2	21.56	32.6	35.71
M25	12.50%	3	23.64	36.6	39.26
M25	12.50%	4	22.07	35.41	40.89
M25	12.50%	5	22.22	35.41	42.38
M25	25%	1	17.92	31.41	35.12
M25	25%	2	20.74	32.23	37.04
M25	25%	3	21.48	31.56	38.07

M25	25%	4	22.37	33.04	39.70
M25	25%	5	21.34	35.56	40.15
M25	37.50%	1	18.96	31.26	31.7
M25	37.50%	2	23.41	26.52	31.41
M25	37.50%	3	18.82	29.04	32.45
M25	37.50%	4	19.7	29.78	33.78
M25	37.50%	5	21.34	33.63	34.52
M30	12.50%	1	29.03	32.15	35.7
M30	12.50%	2	30.67	35.7	37.49
M30	12.50%	3	29.78	33.64	43.26
M30	12.50%	4	30.08	35.11	44.30
M30	12.50%	5	31.41	41.78	43.85
M30	25%	1	27.7	34.97	40.59
M30	25%	2	30.22	32.89	41.03
M30	25%	3	29.48	36.89	42.97
M30	25%	4	31.11	38.38	43.49
M30	25%	5	31.56	39.71	43.97
M30	37.50%	1	20.74	28.75	40.00
M30	37.50%	2	20.74	29.19	41.04
M30	37.50%	3	22.52	32.45	41.18
M30	37.50%	4	21.34	34.82	42.22
M30	37.50%	5	28	34.23	41.93

E. **Data Analysis:** The data containing information of cement content, fine aggregate, water content, fineness of fly ash and compressive strength at 7, 28 and 90 days of age was compiled. Similar data available in various literatures was added to this data. The Regression analysis was carried out treating compressive strength at various ages as dependent variable and other ingredients including fineness of fly ash as independent variables. The regression equations are presented below:

$$\mathbf{y_7 = 0.109\, x_1 + 0.086\, x_2 - 0.289\, x_3 + 0.008\, x_4 + 31.264 \quad (R^2 = 0.687)}$$

$$\mathbf{y_{28} = 0.145\, x_1 + 0.133\, x_2 - 0.429\, x_3 + 0.015\, x_4 + 46.603 \quad (R^2 = 0.694)}$$

$$\mathbf{y_{28} = 0.112\, x_1 + 0.093\, x_2 - 0.640\, x_3 + 0.010\, x_4 + 103.47 \quad (R^2 = 0.779)}$$

Where, $\mathbf{y_7}$ (Comp. Strength at 7 days of age), x_1 = Cement, x_2= Fly ash, x_3= Water quantity in liters and x_4= Blain's fineness of Fly ash.

3. Results and Discussion

A. Test results of physical and chemical properties of fly ash indicate that,

a) The specific gravity of fly ash obtained from hopper no. 1 is least one (2.04). It gradually increases and maximum specific gravity is found for hopper no. 5 (2.28).
b) Initial and Final setting time for coarser fly ash is on higher side as compared to finer fly ash.
c) Fineness when tested by Blain Air Permeability apparatus increases from 1st hopper to 5th hopper.
d) Water requirement for maintaining flow as specified in IS 1727 is higher for coarser fly ash.
e) There is no significant difference, which can be attributed for compressive strength development in relevance to chemical properties

B. Based on the data of table 4, various important graphs are presented in figure 1 to figure 3.

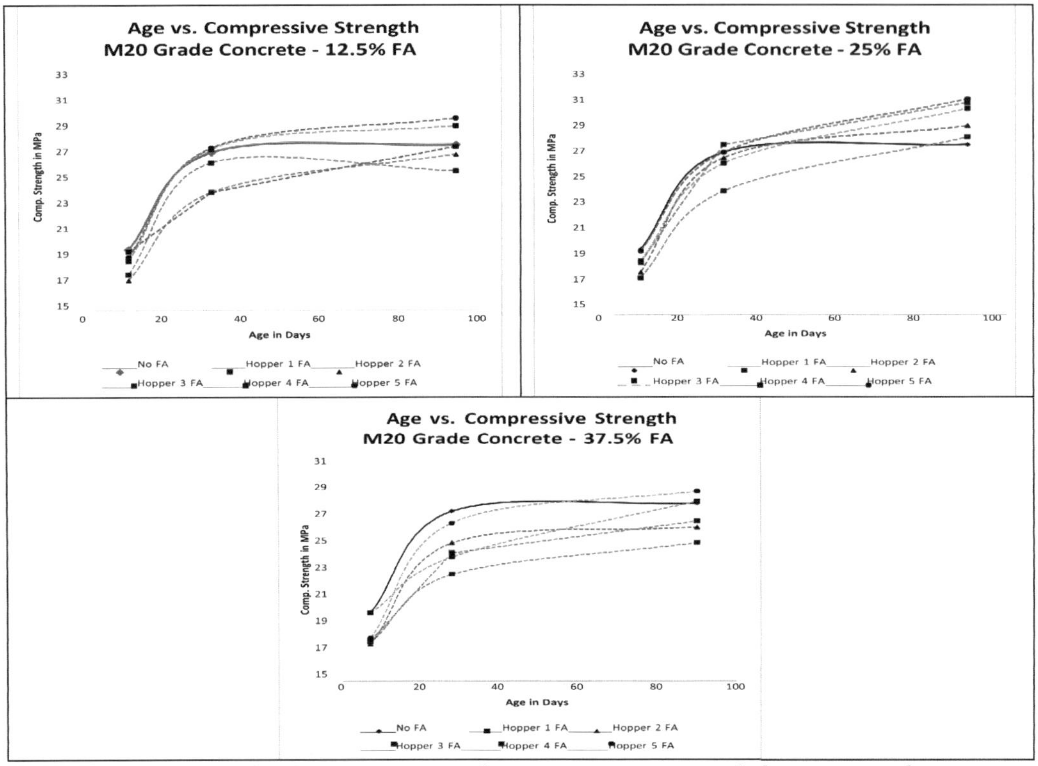

Figure 1. Relationship between age and compressive strength for M20 Grade concrete

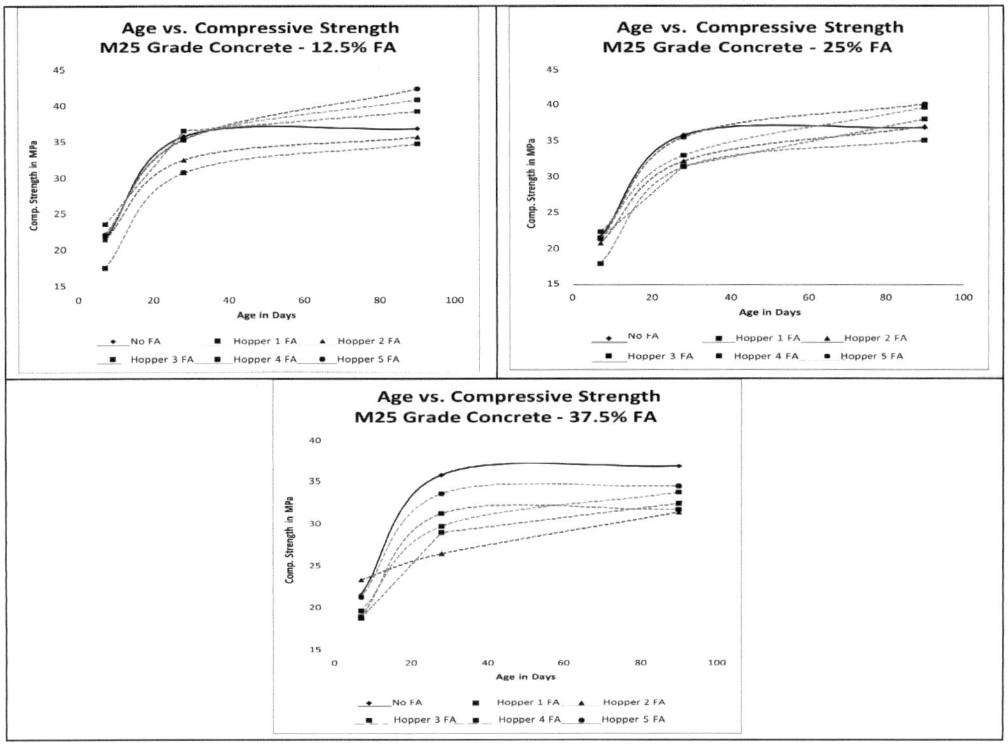

Figure 2. Relationship between age and compressive strength for M25 Grade concret

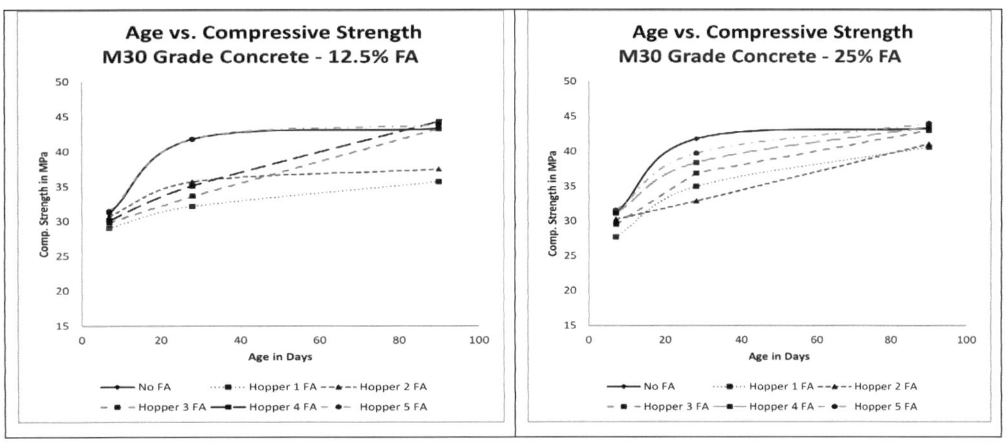

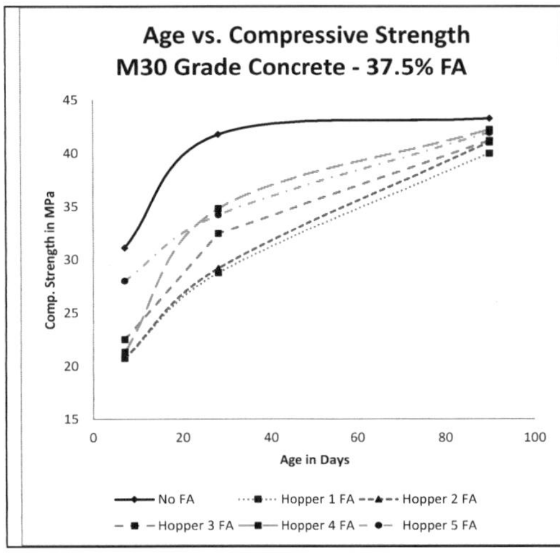

Figure 3. Relationship between age and compressive strength for M30 Grade concrete

From the figure 1 to figure 3, it observed that,

a) In M20 grade concrete, performance of coarser fly ash is relatively lower than finer fly ash samples at all levels of replacement. At 25% replacement of cement by fly ash, it is observed that, fly ash from all hoppers have performed better than plain concrete at 90 days of age. At 37.5 % replacement level, the performance of coarser fly ash is less than plain concrete at all days of age.
b) Similar trends have been observed in case of M25 grade concrete.
c) In M30 grade concrete, the 12.5 % replacement of cement by coarser fly ash, 90 days strength is significantly lower than plain concrete. At 37.5 % replacement level, the performance of all samples is considerably lower than plain concrete at 7 & 28 days of ages. However, it is considerably close to plain concrete.
d) For all ages of concrete, the compressive strength of fly ash concrete increases as the percentage of fly ash increases upto 25% and the downward trend is observed beyond 25% replacement level for all hoppers.

C. **Regression analysis:** Validation of regression equations is presented in Fig. 4.

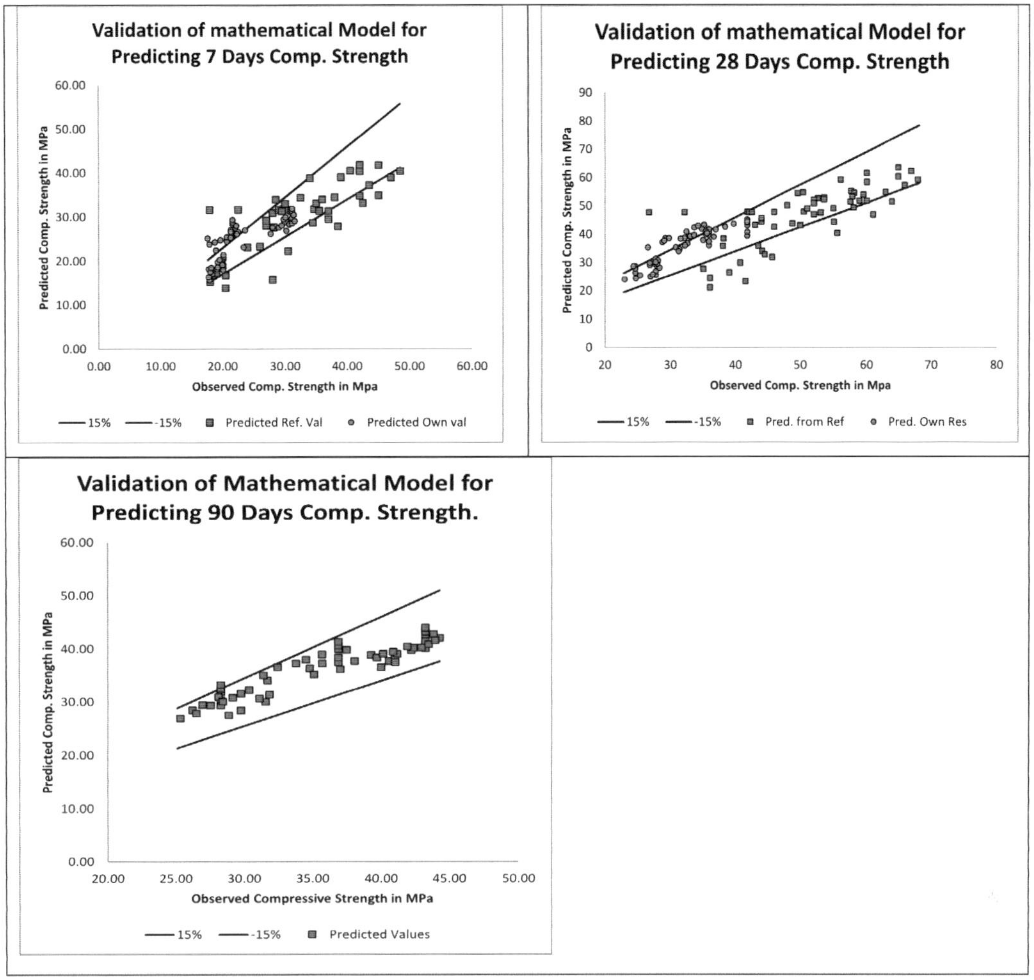

Figure 4. Validation of mathematical model for predicting compressive strength at various age of concrete

From fig. 4, it is observed that majority of predicted values lie within □□15 % of the actual values. This also strengthen that the fineness of fly ash has significant contribution in the development of compressive strength.

4. Concluding Remarks

1. Coarser fly ash is abundantly available at all Thermal Power Stations. However the restrictions impose by both the parts of IS 3812 limits its use. Codal provisions in Japan allow cautions use of all grades of fly ash in various applications.
2. Though the water requirement, when tested as per IS 1727 (Flow Test), the coarser fly ash requires more water. But when tested in concrete with the same w / c ratio for the finer ash concrete, there is no significant loss of slump.
3. The relevance of chemical properties on compressive strength development is not found.

4. The compressive strength of fly ash concrete increases as the percentage of fly ash increases upto 25% and the downward trend is observed beyond 25% replacement level for all hoppers.
5. Compressive strength of concrete containing the coarser fly ash is fairly close to that of finer ones. It is recommended that the coarser fly ash can very well be used in concrete upto 25% replacement of cement with appropriate quality control measures.

5. References

[1] Bhanumathidas N and Kalidas N 2005, "Fly Ash Characterization for use as a Pozzolanic Material", Proceedings of Second International Symposium on " Concrete Technology for Sustainable Development" Hyderabad Feb. 27- Mar. 3 2005, pp 151-164.

[2] Kalgal M R et.al., (2003), *Characterization of RTPS Fly Ash and Pond Ash for use as Partial Replacement to Cement,* Proceedings of "Innovative World of Concrete" Pune, India, Sept. 2003

[3] Kraiwood Kiattikomol and Chai Jaturapittakul, (2001), *A study of ground coarse fly ashes with different finenesses from various sources as pozzolonic materials*, Cement & Concrete Composites, Vol. 23, pp 335-343

[4] Sear L. K. A., (2001), *UK Practice - A Review of Fly Ashes for use in Concrete,* Workshop on Novel Products from Combustion Residues, Morelia ,Spain

[5] Nath P and Sarker P K, (2013), *Effect of Mixture Proportions on the Drying Shrinkage and Permeation Properties of High Strength Concrete Containing Class F Fly Ash*, KSCE Journal of Civil Engineering,
Vol. 17(6), pp 1437-1445

[6] Sumrerng Rukzon and Prinya Chindaprasirt (2008), *Development of classified fly ash as a pozzolonic material*, Journal of Applied Sciences Vol.8 (6) pp 1097-1102

[7] Osman Gencel et.al., (2012), *Combined effects of fly ash and waste ferrochromium on properties of concrete*, Construction and Building Materials, Vol. 29, pp 633-640

[8] Ahmaruzzaman M., (2010), A review on the utilization of fly ash, Progress in Energy and Combustion Science, Vol. 36, pp 327-363

[9] Nagataki S et al 2001, "Classification of Fly Ash According Japan Industrial Standard, JIS A 6201-1999", SP 199- Vol- I, Seventh CANMET/ACI International Conference on Fly Ash, Silica Fume, Slag and Natural Pozzolans in Concrete, pp 29-41, American Concrete Institute

[10] Sivasundaram V., (2004), *Evaluation of Indian fly ashes for the HVFA concrete, Part II: Behaviour in concrete,* The Indian Concrete Journal, Vol 78, Nov. 2004, No 11, pp 41-50

[11] Fournier B. et.al., (2004), *Evaluation of Indian fly ashes for use in HVFA concrete, Part I: Characterisation,* The Indian Concrete Journal, Vol 78, Nov. 2004, No 11, pp22-30

[12] Pande A M and Gupta L M 2007, "Properties of Concrete Ingredients and Their Significance in Compressive Strength Development", Indian Concrete Journal, Vol. 81, No. 6 , June 2007, pp 15-27

[13] Chai Jaturapitakkul et.al., (1999), *A Study of Strength Activity Index of Ground Coarse Fly Ash with Portland Cement, Science Asia, Vol.* 25, pp 223-229

[14] Siddique Rafat, (2004), *Performance characteristics of high-volume Class F fly ash concrete*, Cement and Concrete Research, Vol. 34, pp 487–493

[15] Indian Standard Code of Practice for Pulverized Fuel Ash – Specification Part 1 for use as

IOP Conf. Series: Materials Science and Engineering 1170 (2021) 012006 doi:10.1088/1757-899X/1170/1/012006

pozzolana in cement mortar and concrete. IS 3812 (Part 1) :2003, Bureau of Indian Standards, New Delhi

[16] Indian Standard Code of Practice for Pulverized Fuel Ash – Specification Part 1 for use as admixture in cement mortar and concrete. IS 3812 (Part 2) :2003, Bureau of Indian Standards, New Delhi

[17] IS 1727:1967 (Reaffirmed 1999) Methods of test for pozzolanic materials, Bureau of Indian Standards, New Delhi

[18] Recommended Guidelines for concrete mix Design. IS 10262-2019, Bureau of Indian standards, New Delhi.

[19] Specification for Coarse and fine aggregates from Natural sources for concrete. IS 383-2016, Bureau of Indian standards, New Delhi.

[20] IS:516-1959 (Reaffirmed 2018) Methods of Tests for Strength of Concrete, Bureau of Indian Standards, New Delhi

The Growing Use of Nanotechnology in the Built Environment: *A Review*

M Ghosal[1,2] and A Chakraborty[3]

[1] Jt. Secretary, Coal Ash Institute of India

[2]PhD Scholar, IIEST, Shibpur, India

[3]Associate Professor, IIEST, Shibpur, India

E-mail: mainakghosal2010@gmail.com

Abstract. Though the past industrial revolutions had unleashed many of the world's current problems including the Covid-19 virus but the 4th industrial revolution promises a sustainable future through the use of advanced materials like Nanotechnology. Not at all new, the use of this disruptive technology had its footprints in the pages of history - Ajanta Paintings, Damascus Sword & many others. Nano Titanium dioxide (58%) and Nano silica (21%) are the most used advanced materials in the nanotechnology-based building products while China in Asia and Germany in Europe are the two leading countries in the field of production of nanomaterials for building industry. India has been amongst the five top countries globally for technical publications in nano sciences & technology. This paper delves into the world of advanced nanomaterials and studies their impact on the construction industry vis-a-vis the built environment in India and abroad.

Keywords – Country, Industry, Nano, Nanotechnology, Revolution

1.0 Introduction

Rapid industrialization after 1800s has led to many of the world's current environmental problems like, over-crowding, climate change, unsafe levels of pollution and toxins in the air, water and soil and consequently into food chain, the depletion of fishing stocks, high mounds of waste on land and in large areas in the ocean, loss of biodiversity, and deforestation. However, the 4th Industrial revolution or Industry 4.0 promises a sustainable future for all by leapfrogging traditional developments and ushering in fastest-ever technological disruptions like artificial intelligence (AI), robotics, advanced materials (including nanomaterials), 3D printing, blockchains and biotechnology, the Internet of things (IoT), autonomous vehicles and drones. Though 3D printing, robotics, blockchains and drones are already in use or close to deployment others like advanced materials, autonomous vehicles, biotechnologies, AI, etc. are in the various stages of development and testing. However, the question raised in the World Economic Forum in Davos was that whether we ready for this type of Technological revolution? This paper delves into the world of advanced nanomaterials and studies their impact on the construction industry vis-à-vis the built environment in India and abroad.

Fig.1: Energy Efficient House of the Future [1]

There are many applications of nanotechnology in building construction. It has given a new dimension for construction materials with its unique and amazing properties._However, this technology is not at all new and has its footprints in Ajanta Paintings, Tipu Sultan's Damascus Sword, and others. As per existing literatures, among the most important nanomaterials that are found in the building industry are nano-titania (TiO_2), silver, nano-zinc oxide (ZnO), nano-silica (SiO_2), calcium silicate, and aluminum phosphate and carbon nanotubes (CNT).

Titanium dioxide (TiO2) is found in 58% of all nanotechnology products in the field of building. Increasing mechanical properties of the samples in the main structures, waterproofing application on internal and external surface of the buildings, prevention of dirt, UV-resistance, production of strengthened, self-repair and self-cleaning concrete, fireproof and self-cleaning glass and energy saving glass are the most important applications of these nanoparticles in building industry [5].

Nano silica (SiO2) possesses a high share of nanotechnology-based building products by having a share of 21%. The use of nano silica decreases the viscosity of cement mortar which results in a decrease in water consumption. Based on studies, silicone chains are created in the nanometric structure of cement with various lengths. The length of the chains is very important in the stability of cement. Fireproof glass is another achievement of nanotechnology in the field of building industry. This product is produced by adding a transparent layer of silica nanoparticles between the two sheets of glass. When the glass becomes warm, the transparent layer turns into a hard, dark and fireproof coating [5].

The tasks of architects and engineers are now made much easier through nanotechnology by introducing construction materials such as seismic wallpaper, thermal insulation and additives which are easy to use and can aid in improving the durability, and structural integrity of our infrastructure. Amongst Asia and in Europe, China and Germany are the two leading countries in the production of nanotechnology-based products in the field of building industry; according to the statistics published by Statnano Database website. They are said to have possess a share of 30% and 25% of all nanotechnology-based building products in the global market, with Shanghai huzheng Co. alone producing 44 products. It is followed by USA (19%), Switzerland (6%), Greece

(5%), Poland (5%), Singapore (3%), Finland (2%) and Denmark (1%). Surprisingly, India the 5th largest economy and 3rd in terms of GDP, stands nowhere in either active nano technological research or product development.

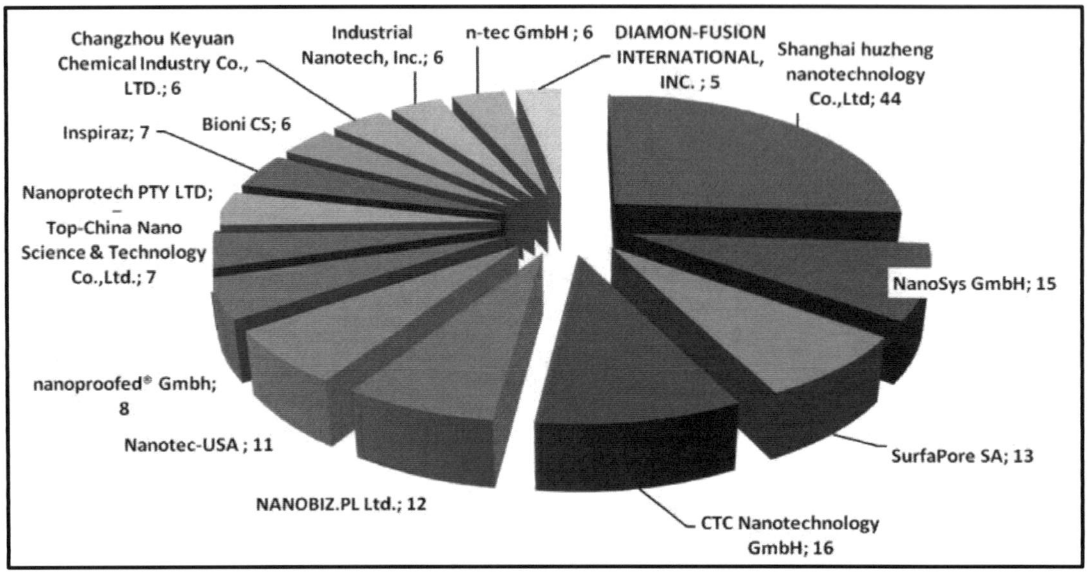

Fig.2: Cluster of Nanotechnology-Based Building Products' Enterprises [5]

2.0 The Global Development Story of Nanotechnology

The manifestations of globalization are many for example - bank building's and company head quarter's sky scrapers, standardized hotel chains, franchise restaurants, shopping malls, branded stores and boutiques, specialty & multi-specialty hospitals, theme parks, fitness centers and multi-plex cinemas. The predominance of load bearing structures of the 19th century gave away to reinforced concrete construction (RCC) structures which dominated the whole of 20th century with some steel structures. Globalization has brought along new architectural trends which incorporate sustainability and the role of technology in it. With its own buzzwords, nanotechnology is a sort of the miniature rebellion that was initiated with the discovery of sophisticated microscopes in the later part of the 20th century. Because of the opportunity it provides in creating new features in wide ranging applications, nanotechnology is attracting more funding than any other technological areas with an unbeatable patent processing record but the problem with it is that it requires lengthy team-work with people from multi-disciplinary fields like physics, chemistry, biology, engineering and medicine working together.

Technology plays an important role in globalization thus making our world a smaller and better place to live in. With the advent of 21st century, nanotechnology or manipulation of matter at the atomic level made a leap from science fiction (Star Trek, Spiderman, etc.) to reality, with both public and private players around the globe began focusing on its projected economic benefits. One expected impact is that nanotechnology may augment and magnify the already existing discrepancy between the rich and the poor. On the other hand, it may also reduce the disparity by making things more affordable over a period of time. It is contemplated that like the digital divide that has mushroomed with the introduction of ICT and computers, it is very likely that with differential rates of diffusion of these technologies, there will be 'nano divide'- nano *have* & *have not* countries. Some advocates of nanotechnology however, argue that scientific advances in this nano-scale will bring a prosperity to material scarcity like wood or oil, due to the ability of the synthesizing power of nanotechnology. Discoveries in this domain would also lead to intense competition in acquiring patents and intellectual property rights amongst the nations reinforcing the concept of liberalization with MNCs continuing to play a promoting role in addressing global policies in the matter of trade, relations and policy.

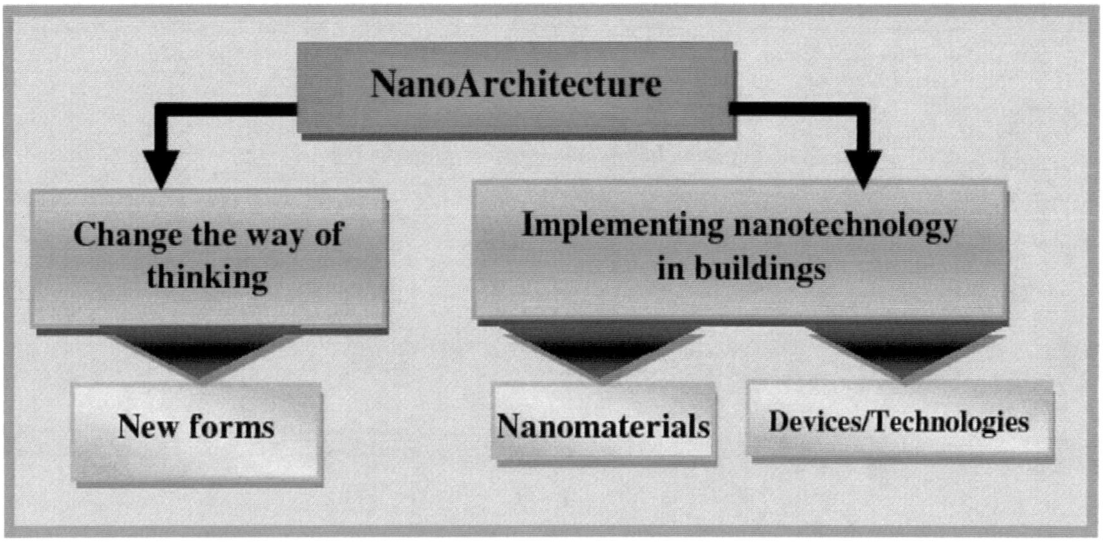

Fig.3: Nanoarchitecture: Nanotechnology in Buildings [2]

There are numerous examples of nanotechnology products moving from simple to complex. From the simple dirt repellant nano-coatings we have moved towards mobile phones which dramatically have become smaller & smaller but cleverer, faster & cheaper. The applications of nanotechnology are numerous and can be found in medical science, textiles, mobile communications, transport, architecture and engineering. Nano functions can be effectively used in the interiors of commercial buildings as shown in Fig.4 for a Hotel Room or Office Room or also Hospital Room but strategically after studying their cost-benefit analysis.

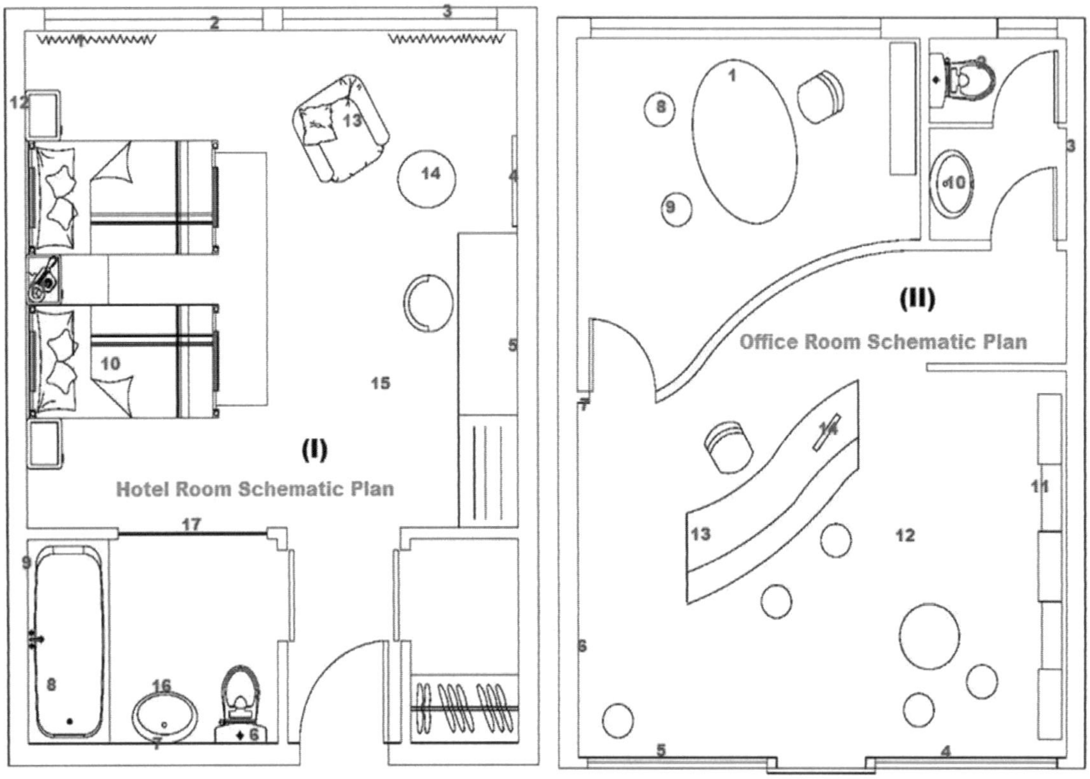

Fig.4: Schematic Plans of Hotel & Office Room for strategic use of Nanomaterials

Table 1. Nano-enabled products for different commercial usages

(I) For Hotel Room	(II) For Office Room
01- Curtains: Air-purifying	01-Glass Table: Anti-fingerprints
02- Windows: Self cleaning(photocatalytic)	02-W.C.: Easy to clean
03- Windows: Self cleaning(photochromatic or electrochromic)	03-Walls: Nanoparticles ceramic covering
04- TV: Anti reflective	04-Windows: Self cleaning (photochromatic or electrochromic)
05- Wall Paint: Air-purifying	05-Windows: Self cleaning (photocatalytic)
06- W.C.: Easy to clean	06- Walls: Nanoparticles ceramic covering
07- Mirror: Anti-fogging	07-Switches & Handles: Anti-bacterial, non-stick
08- Bath tub & Shower screen: Easy to clean, non-stick	08-Chairs: Dirt-repellant
09- Walls: Nanoparticles ceramic covering	09-Upholstery: Air-purifying
10- Bedding: Anti-bacterial	10- Sanitaryware: Anti-fingerprints
11- Light Switches: Anti-bacterial, non-stick	11-Screen:Anti-reflective
12- Wall Paint: Air-purifying	12-Carpet: Air-purifying
13- Upholstery: Air-purifying	13-Counter: Anti-fingerprints
14- Glass Table: Anti-fingerprints	14-Screen: Anti-reflective

15- Carpet: Air-purifying	----
16- Sanitaryware: Anti-fingerprints	----
17- Frosted Glass: Anti-fingerprints	

With the increasing availability of nanotechnological consumer products worldwide, it was a general expectation that the public at large would be rather familiar with the term 'nanotechnology' and understand what it means. Though companies using nanotechnology often deny using them to their end-customers and thus in the midst of this accelerating commercialization, public worldwide remain largely in dark about nanotechnology [8] perhaps due to fear of things they can't visualize or perceive. Major studies in USA, Europe, Canada and even China have shown that majority of public has heard little or nothing about nanotechnology. What is even more surprising is that they have little or no faith in their government's ability to manage the potential risks posed by nanotechnology. See what havoc nanoparticles like *Coronavirus* are playing in the world today with governments as mute spectators.

Table 2. Nano-enabled construction products identified in the UK [3]

Type	Number	Nanomaterials commonly used (bracketed entries less common)
Coatings	70	Silica, titanium, silver (CNTs)
Glass	23	Metal oxides
Concrete	22	Silica, titanium (CNTs)
Steel	11	Nanostructured
Insulation	10	Silica aerogels
Composites	1	(CNTs, nanoclays)
Other(roofs, floors)	19	Titanium, aluminum

3.0 Nanotechnology: India's Story

The micro-electronics revolution which occurred during 1970-80s failed to take off in India due to its then underdeveloped science and technological scenario. After the advent of liberalization-privatization-globalization in 1990s, the nation came to know of this nanotechnology but the government took no interest to develop it. However, in 2001, Nano Science & Technology Initiative (NSTI) was started under the Department of Science and Technology (DST) of the Ministry of Science as the nodal agency of the with a meager initial budget of only Rs.6 crores in the 11th - 5 Year Plan though the budget outlay for DST was Rs.20,000 crores. This was when private organizations like Reliance, Tata and a few other players had already made more investments in nanotechnological research.

In 2004, under the then President Dr. A.P.J.Abdul Kalam, a workforce was formed and in 2007 a Nano Mission was established in India with a budget of Rs.1,000 Cr. with 19 Centers of Excellence. That 'nano-mission' has been working to help scientists, institutions-industry interface in terms of promoting basic research, development of adequate manpower resources, international collaborations, augmentation of the infrastructure for research and generation of socially useful products. This nano-mission has resulted in about 5000 research papers and some useful products like nano hydrogel-based eye drops, pesticide removal technology for drinking water, water filters for arsenic and fluoride removal and nano silver based antimicrobial textile coating.

India has been able to rank amongst the top five countries in the world for Scientific Publications in Nanoscience & Technology due to the efforts led by the Nano Mission. The Nano Mission has established national dialogues to promote R&D in the development of standards for nanotechnology and for laying down a National Regulatory Framework Road-Map for Nanotechnology (NRFR-Nanotech). Recently in the pandemic times, the Department of Science and Technology and the Science and Engineering Research Board (SERB) called for a Short-term Research Grant for Antiviral Nano Coating COVID-19 in April 2020 to combat the pandemic.

The budget for Nano Mission which has decreased considerably from Rs.1,000 crores in 2007 to Rs.650 crores in 2012 to NIL now [9]. But it developed physical infrastructure, with the DST, establishing Units on Nano Science, Centers for Nanotechnology, Centre of Computational Materials Science and Thematic Units of Excellence on Computational Material Science in various universities/, IITs and government research centers/ laboratories across the country. As far as human resource development is concerned, the Nano Mission has launched PG programmes (M.Sc. and M. Tech in Nano science and Nanotechnology) at various universities and colleges all across India. The **NITI Aayog** which replaced the Planning Commission is completely silent on nanotechnology, except in one area, as expressed in its strategy paper, **New India @ 75** which says that - Foreign collaborators, consultants, visiting faculty, adjunct scientists, etc., need to be involved in pursuing R&D in the emerging areas of basic sciences such as **nano-technology**, etc. DST, in collaboration with Indian Missions abroad, may identify discipline wise foreign experts who can collaborate with Indian scientists to take basic research in these areas to the next level.

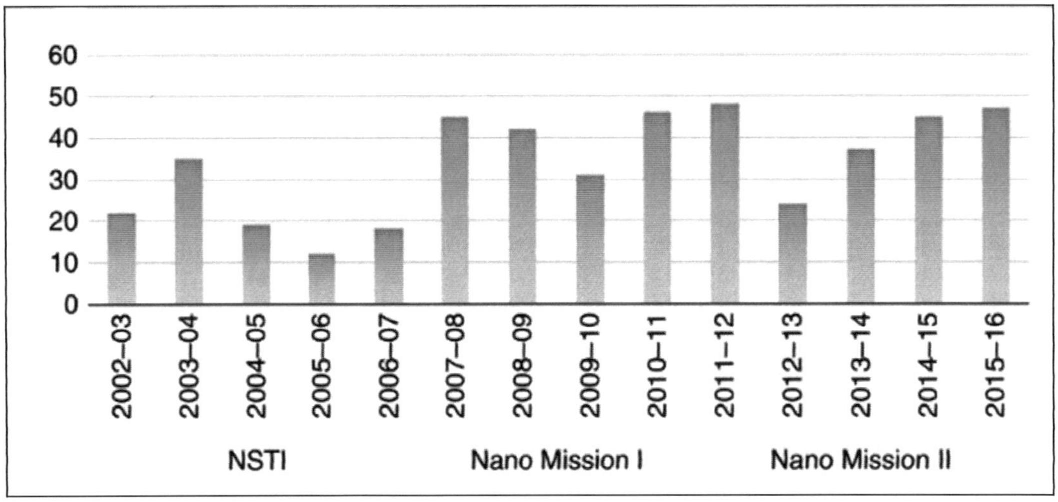

Fig.5: Government-led nanotechnology projects sanctioned in India since 2002[4]

4.0 Conclusions

Globalization cannot be simply stopped and countries have jumped from being developing to being developed in the past by powering a manufacturing/ industrial revolution albeit remaining self-sufficient of the food front. Despite having a huge labor force and initiating economic reforms in 1991, India seems to have missed the last bus of manufacturing revolution. As against the conventional wisdom, India appears to have shifted from farming to services without having gone through an industrial stage which contradicts the experience of other fast-growing Asian economies like China. The Asian counterparts (China, Japan, Taiwan, South Korea and countries of South East Asia) essentially followed an export-oriented manufacturing strategy focused on low-end products which later progressed up the value chain to create economic growth. However, after the 1991 economic reforms, India started to export complex automobile parts and pharmaceuticals along with information technology and became a global hub of the business process outsourcing industry. Later, India cautiously experienced a huge expansion in banking, hotels, airlines, cable television, telecom etc. but none of this was low-end, like was the case of

Asian countries. Hence, it followed the high-end model of services, without going through an industrial/ manufacturing stage which saw the country's largest car-maker Tata failing to increase the sale of Nano and ultimately stop its production though it did very well in the foreign countries.

As far as nanotechnology is concerned there are hardly any codes or standards which holistically address the performance evaluation of nanomaterials in cement concrete apart from the recent report from American Concrete Institute (ACI 241R-17) in 2017[7]. Though the revised National Building Code of India, 2016 addresses nanotechnology/ nanomaterials as sustainable and as alternate building materials. There exists a lack of coherence between the academic field and the industrial requirements and the need for a strategy in the new industry creation in the nanotechnology field, where not only public research institutes, but also existing industries (including SMEs) can participate [7].

As seen in countries like Japan, China and Europe, to nullify the size disadvantage of nanomaterials and overcome diseconomies of scale, SMEs often utilize the process of technology transfer from public funded R&D institutions [6,7].

Indian codes such as IS:456-2000 and the revised IS:1026-2019 mention Silica fume (with more than 70% having particle size diameter within 100nm range, but they are not marketed as nanomaterials by the manufacturers) for concrete but which has been in use for over 30 years and is widely available in the products of most major global concrete suppliers. Hence, commercialization of nano-products still remains a challenge. As happens with any emerging technology the parameters of Strategy + Demand + Policies + Initiatives are needed in sequence but in India does not have any regulatory framework for nanotechnology till now.

The pandemic has given India an opportunity to get on the bus for microelectronics and nanotechnology to boost 'Make in India' a success.

References

[1] https://aecom.com/wp-content/uploads/2015/10/Blue-Book-2014.pdf

[2] A Y Mohamed. 2015. 'Nano-Innovation in Construction, A New Era of Sustainability', Conference Proceedings.

[3] Jones, W. et al., 2016. 'Nanomaterials in construction – what is being used, and where?' Proceedings of the Institution of Civil Engineers: Construction Materials, 172 (2), pp.49-62.

[4] 'The Emerging trends in agri-nanotechnology: fundamental & applied aspects' (2018)/ editors: Prof. H. B. Singh, Dr. Sandhya Mishra, Prof. Leonardo Fernandes Fraceto, Dr. Renata de Lima, Published by CAB International, USA, ISBN-13:9781786391445.

[5] https://statnano.com/news/50098/Growing-Application-of-Commercial-Nanotechnology-based-Products-in-Construction-Industrial-Sector

[6] Purushotham, H., 'Transfer of nanotechnologies from R&D institutions to SMEs in India', *Asia-Pacific Tech Monitor*, vol. 29, no.4, pp. 23-33, October-December2012.

[7] Ghosal M., Chakraborty A.K. (2021) A Study on Performance of Carbon-Based Nano-enabled Cement Composites and Concrete. In: Ashish D.K., de Brito J., Sharma S.K. (eds) 3rd International Conference on Innovative Technologies for Clean and Sustainable Development. ITCSD 2020. RILEM Book series, vol 29. Springer, Cham. https://doi.org/10.1007/978-3-030-51485-3_29.

[8] Allhoff, Fritz and Patrick Lin (eds). 2008. *Nanotechnology and Society: Current and Emerging Ethical Issues* (Dordrecht: Springer).

[9] Kumar, Amit & Desai, Pranav. (2014). Mapping the Indian nanotechnology innovation system. World Journal of Science. 11. 10.1108/WJSTSD-09-2013-0039.

Multi -Response optimization of Electric Discharge Machining (EDM) process parameter for Aluminum based Hybrid Metal Matrix Composite using GRA

Saurabh shukla[1], D.K Singh[2]
PG Student[1], Department of Mechanical Engineering,
Professor[2], Department of Mechanical Engineering,
[1,2]Madan Mohan Malviya University of Technology Gorakhpur273010,UP,India
[1)]shuklasaurabh893@gmail.com
[2)]dhirenks@gmail.com

Abstract Aluminum based hybrid metal composite (MMC) have been applied to many various applications in the area of automobiles and aerospace. This research article shows the multi-response optimization of EDM of process parameter for Aluminum based metal matrix hybrid composite (Aluminum 6061/Al_2O/TiO_2).The optimum process parameter of EDM as like pulse on time (Ton), current (Ip), duty cycle (t), voltage (V) on material removal rate and surface roughness were investigated. Hybrid Aluminum metal matrix composite was machined by using tool which is made of copper materials, of ϕ12 mm for experimentation. The mechanical stirring procedure is used to fabricate the hybrid MMC. The design of experiments was conduct through the Response Surface Methodology (RSM).The experimental result declares that the main influencing input parameter is current.

Key Words: *EDM, Metal Matrix Composite,MRR, Surface roughness, GRA,*

1. Introduction- In conventional machining process machining of MMCs is very difficult due to high tool wear rate, less production, poor economy of get accuracy from the new method of material removal which is different from conventional machining process as compared to non-traditional machining (NTM) processes as like EDM, ultrasonic machining (USM), Abrasive Jet Machining(AJM), Abrasive water jet machining (AWJM), Laser Beam Machining(LBM), from the above EDM is the most widely useable to machine MMCs, because these material are hard, high strength, temperature resistance and electric conductive material [1,2]. V. Suresh et.al [3] attempt to fabricate the hybrid MMC using stir casting technique with Aluminum alloy (LM25) reinforced with graphite (Gr), boron carbide (B_4C) and analyze the machining parameter such as material removal rate, surface roughness, tool wear rate using wire EDM. The result shows that the most influencing input parameter is current, which affect directly on removal of material. The maximum removal of material is obtained from the combination of current 7, pulse duration 4μs, pulse of time 6μs and less influencing parameter for MRR is pulse duration and pulse off. M. Mastana et.al [4] investigate the machining parameter of EDM on aluminum alloy AA6063.The result shows that material removal rate of AA6063 increased with increases of pulse on time and pulse of time and decrease with duty factor and gap voltage. S. Manish et. al [5] investigated that machining parameter MRR, tool wear rate, radial overcut, flatness of Aluminum LM-6/SiC/B_4C hybrid composite fabricated by stir casting method and result show that the current is the most influencing parameter for affecting the multiple performance. K. Alip et.al [6] in this paper analyze the tribological and microstructure behavior of aluminum alloy reinforced with SiC/TiC hybrid metal MMC and result show that higher reinforcement content increases the microstructure of aluminum composite. Hardness increased with increment of reinforcement material and also wear rate of composite material increase with increase in load. Natryan et.al [7] optimization of hybrid metal matrix composite of aluminum AA606/Al_2O_3/SiC was done by using the squeeze casting method and Taguchi based method was used to analyze and optimize the process parameter. The optimized result shows that A2, B2, C2 such as load 20N (level), sliding velocity1m/s(level) and sliding distance 1200m (level) and respectively. K. Arun et.al [8] light weight as well as light weight is the most widely required property for the advance industries. Research is ongoing to address the difficulties identified in the growth of metal matrix composite. Yadv. R et.al [9] this paper shows main focus is on the tribological behaviour of aluminium based MMC fabricated by stir casting process. The result shows that mechanical properties increased with increment of reinforcement particle and improve hardness and reduce ductility and also in addition of silicon carbide in aluminium based MMCs increase in the hardness and strength and fly ash increase the wear rate of MMC, this paper show that very less research are done in the area of organic reinforcement.

2. Material and Method

2.1 Material

The AL6061 is used as a matrix material with feature of hardened alloy exhibits better mechanical and wear properties. It is most widely used alloy in area of automobile sector and also in aerospace [10]. It has included with high melting element which is magnesium and silicon. The chemical and mechanical properties of matrix and reinforced material are shown in Table 1 & Table 2.The reinforcement Al_2O_3 & TiO_2 of the average particle size 20 μm respectively used in aluminum matrix composite. The cooper is most widely and appropriate material use for machining of aluminum-based metal matrix composite. Copper has better electrical and thermal properties which are fulfill the requirement of machining of composite material as well as for Electric Discharge Machining [11-12].

Table 1. Mechanical Properties of Aluminum6061 [13]

Density (g/cm³)	Strength (MPa)	Modulus (GPa)	Fracture Toughness (MPa)m $^{1/2}$	Coefficient of thermal expansion
2.7	180	75	30	23.8

Table 2. Properties of $Al_2O_3$and TiO_2 [14]

	Density (g/cm³)	Melting point (°C)	Boiling Point (°C)	Thermal conductivity(W/m.k)	Strength (MPa)
Al_2O_3	3.9	2040	2977	6.9	2000 -4000
TiO_2	4.23	1843	2972	20-30	333.3

2.2 Fabrication of composite material using stir casting method

The work material was fabricated by stir casting method. The composition of fabricated specimen is equal weight for aluminum oxide (5%) and titanium oxide (5%) with metal matrix composite alloy AA6061(90%).The Aluminum 6061 was heated at 660 °C in muffle furnace to melt it completely [15] .The Al_2O_3 and TiO_2 reinforced particle were preheated at 500 °C to reduce the oxide on the surface. The preheated reinforcement particle was added in molten metal slowly and mixing performed through the mechanical stirrer approximate 10 to 20 minute to uniformly distribution of matrix [16]. Finally, the molted metal poured in to the mould dimension of 75mm*55mm*10mm and left for cooling at room temperature and figure 1 (a) show the workpiece after machining. The experiment was performed on EDM (SMART-ZNC) powered with 3 phase servo stabilizers. There are 27 experiment runs were performed for machining of AL 6061 composite material.

(a)

Figure 1 (a) Workpiece after Machining

2.3 Machin

Table 3. Experimental parameter

Process Parameter	Level		
	Low	Medium	High
Pulse current (Ip)	4	6	8
Voltage (V)	40	60	80
Pulse on Time(T_{on})	100	200	300
Duty Cycle (t)	6	7	8

The machining parameter are taken as Pulse on time (Ton), current (Ip), voltage (V), duty cycle (t), So these are taken as low, medium, high level which is different. The machining output parameter is obtained in the form of MRR and surface characteristics. The MRR is calculated to take the weight of workpiece after machining and before machining, and the quality of surface is calculated by the roughness tester device Taylor and Hobson.

MRR calculated by given formula

$$MRR = \frac{Initial\ weight - Final\ weight}{Time}\ gram/min \quad \ldots\ldots\ldots\ldots(1)$$

Table 4. Experimental result

S.No.	Current(Ip)	Voltage(V)	Ton (µm)	Duty cycle (t)	MRR gm/min	Ra (µm)
1	4	40	200	7	0.027	5.50
2	8	40	200	7	0.118	4.00

3	4	80	200	7	0.026	6.30
4	8	80	200	7	0.094	4.40
5	6	60	100	6	0.086	6.40
6	6	60	300	6	0.051	3.90
7	6	60	100	8	0.078	6.90
8	6	60	300	8	0.057	7.00
9	4	60	200	6	0.029	6.80
10	8	60	200	6	0.133	4.60
11	4	60	200	8	0.027	6.30
12	8	60	200	8	0.118	8.40
13	6	40	100	7	0.076	7.10
14	6	80	100	7	0.075	6.60
15	6	40	300	7	0.047	7.50
16	6	80	300	7	0.037	7.90
17	4	60	100	7	0.053	5.60
18	8	60	100	7	0.158	7.80
19	4	60	300	7	0.016	5.20
20	8	60	300	7	0.086	10.60
21	6	40	200	6	0.071	7.70
22	6	80	200	6	0.054	10.00
23	6	40	200	8	0.079	7.20
24	6	80	200	8	0.070	8.60
25	6	60	200	7	0.077	6.80
26	6	60	200	7	0.088	8.40
27	6	60	200	7	0.069	10.40

2.4 Conduct of Experiment

Grey Relational Multi Response Optimization process is used to evaluate experiment. According to the L27 experiment from Response surface methodology, the experiment planned and performed. The following steps are used in grey relational analysis to obtain better value [17-19].

Normalization of MRR and TWR

(a) For smaller the better

$$X_i(k)= \frac{maxXi(k)-Xi(k)}{maxXi(k)-minXi(k)} \qquad (2)$$

Where

Xi (k) = the value of quality characteristic after GR generation

maxXi (k) = maximum value of Xi

minXi (k) = minimum value of Xi

(b) For higher the better

$$Xi\ (k) = \frac{Xi(k) - minXi(k)}{maxXi(k) - minXi(k)} \quad \text{.......................................(3)}$$

Deviation Sequence

Deviation sequence is defined as amount of each response which gains from reference.

$$\Delta_{0i}(k) = X_0(k) - X_i(k) \quad \text{...(4)}$$

$\Delta_{0i}(k)$ = deviation sequence
X_0 (k) = reference sequence =1

Grey relational coefficient (GRC)

The grey relation coefficient is computed to reveal the relation between ideal and actual normalized experimental result.

$$£_i\ (k) = \frac{\Delta \min(k) + £\Delta\max(k)}{\Delta 0i(k) + £\Delta\max(k)} \quad \text{.. (5)}$$

Where

£= distinguish coefficient =0.5 to give equal importance to all value.
Δmin = the lower value of $\Delta_{0i}(k)$
Δmax = the highest value of Δ_{0i} (k)

Grey relational grade (GRG)

$$\gamma_i = \frac{1}{n}\sum_{k=1}^{n} £i(k) \quad \text{..(6)}$$

Where

n = no. of response
k= suggestion for response

Table 5. Normalization of MRR and SR

S.No.	MRR	Ra
1	0.080	0.761
2	0.719	0.985
3	0.070	0.642
4	0.547	0.925
5	0.490	0.627
6	0.246	1.000

7	0.436	0.552
8	0.292	0.537
9	0.093	0.567
10	0.824	0.896
11	0.080	0.642
12	0.720	0.328
13	0.420	0.522
14	0.418	0.597
15	0.221	0.463
16	0.147	0.403
17	0.260	0.746
18	1.000	0.418
19	0.000	0.806
20	0.491	0.000
21	0.389	0.433
22	0.270	0.090
23	0.446	0.507
24	0.380	0.299
25	0.429	0.567
26	0.508	0.328
27	0.373	0.030

Table 6. Deviation sequence for responses

S.No.	MRR	Ra
1	0.920	0.239
2	0.281	0.015
3	0.930	0.358
4	0.453	0.075
5	0.510	0.373
6	0.754	0.000
7	0.564	0.448
8	0.708	0.463
9	0.907	0.433
10	0.176	0.104
11	0.920	0.358
12	0.280	0.672
13	0.580	0.478

14	0.582	0.403
15	0.779	0.537
16	0.853	0.597
17	0.740	0.254
18	0.000	0.582
19	1.000	0.194
20	0.509	1.000
21	0.611	0.567
22	0.730	0.910
23	0.554	0.493
24	0.620	0.701
25	0.571	0.433
26	0.492	0.672
27	0.627	0.970

Table 7. Grey Relational Coefficient (GRC) for response

S.No.	MRR	Ra
1	0.352	0.677
2	0.640	0.971
3	0.350	0.583
4	0.524	0.870
5	0.495	0.573
6	0.399	1.000
7	0.470	0.528
8	0.414	0.519
9	0.355	0.536
10	0.740	0.827
11	0.352	0.583
12	0.641	0.427
13	0.463	0.511
14	0.462	0.554
15	0.391	0.482
16	0.370	0.456
17	0.403	0.663
18	1.000	0.462
19	0.333	0.720
20	0.496	0.333
21	0.450	0.469

22	0.407	0.354
23	0.474	0.504
24	0.446	0.416
25	0.467	0.536
26	0.504	0.427
27	0.444	0.340

Table 8. Grey Relational Grade (GRG)

S.No.	Grey Relational Grade	Rank
1	0.514	10
2	0.806	1
3	0.466	18
4	0.697	5
5	0.534	6
6	0.699	4
7	0.499	13
8	0.467	17
9	0.446	21
10	0.783	2
11	0.467	16
12	0.534	7
13	0.487	15
14	0.508	11
15	0.437	22
16	0.413	25
17	0.533	8
18	0.731	3
19	0.527	9
20	0.414	24
21	0.459	20
22	0.381	27
23	0.489	14
24	0.431	23
25	0.501	12
26	0.465	19
27	0.392	26

From above, Grey Relational Grade using L27 experiment shows that the maximum Grade presents the compatible experimental result is nearer to the ideal normalized value. So in this L27, experiment run no. 2has the better multi performance feature because it has larger Grade.

3. Result and Discussion

Main Effect Plot for MRR

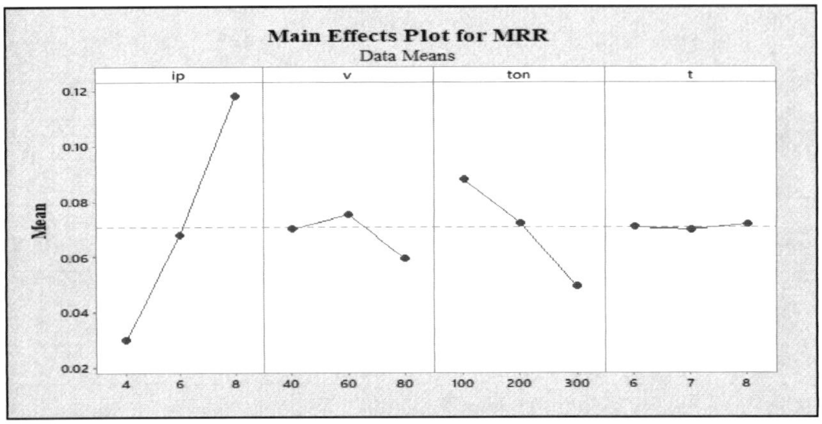

Figure 2. Variation in MRR

From the above main effect plot graph for material removal rate are analyzes in respect of input parameter. At the initial stage for current 4Ip, the lower value of current show the less quantity of material removed but in case of medium value of current at 6 Ip, the quantity of material removal increases as compare to the current 4. The higher value of current shows the higher material removal as comparison of current 4 and 6. From the above analysis of main effect plot graph the result shows that the, when the value of current increases then material removal rate is also increased and also decreased when current decreased. From the analysis of voltage, the result shows that the material removal rate increases between the 40Vto 60V but after that the material removal rate decreases between the 60V to 80V, due to the formation of recast layer.

The above analysis is shows that maximum MRR is obtained at the value of 100μs. With the increases in pulse duration from the 100-300μs, the material removal decreased. The decrement in material removal take place may be due to higher pulse duration more material gets melt between tool and work material, so there is need proper flushing time for debris between tool and work material so arcing happen, this cause of decrement in material removal. The above analysis shows that maximum MRR is obtained at value 6 with increases of duty cycle MRR increases slightly from value at 7-8.

Main Effect plot for Surface Roughness

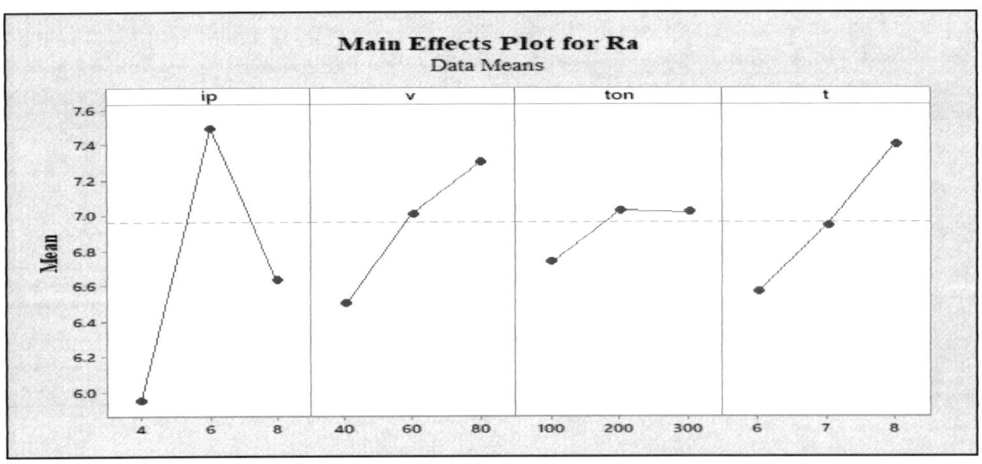

Figure 3. Variation in Surface Roughness

From the above analysis, the surface roughness decreases when current decreases. But surface roughness increases from current value at 4-6Ip, and decreases when current increases from current value 6-8Ip. When the voltage decreases surface roughness also decreases, but slightly increases when voltage increases, minimum surface roughness obtained when pulse on time decreases but increasing in pulse on time the result shows slightly decrement in surface roughness from value 200-300µs.The effect of duty cycle shows that surface roughness is minimum when duty cycle decreases but surface roughness increases with increasing of duty cycle from value at 7-8

Mean calculation for grey relational grade (GRG)

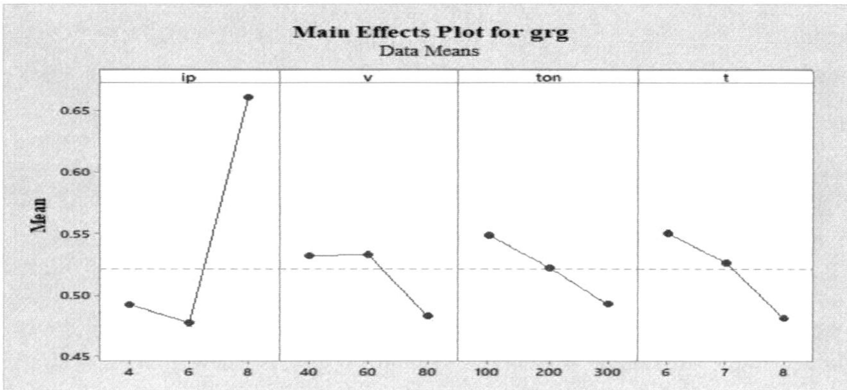

Figure 4 Main GRG effect plot

From the table 9, it defines the information about mean of grey relational grade and it shows the difference between the higher and lower value which decided the rank of each level of machining parameter. The ranking of the process parameter gave the current is most influence parameter on output response and after that duty factor, pulse on time, voltage influence respectively.

Table 9. Response Table for Mean

Parameter	Level 1	Level 2	Level 3	Max-Min	Rank
Current(Ip)	0.496	0.477	0.661	0.183	1
Voltage (V)	0.532	0.533	0.499	0.033	4
Duty Factor (t)	0.550	0.526	0.481	0.069	2
Ton (µs)	0.549	0.522	0.493	0.056	3

4. Conclusion

The stir casting method is very efficient and easy fabrication process to fabricate the hybrid metal matrix composite. In this work the experiment was performed by design of experiment L27 Box-Behnken method. Grey Relational Analysis is used to optimize the EDM process parameter for aluminium based hybrid metal matrix composite. According analysis of response table the current is most influence parameter as compared to other parameter and the combination of other. In this present study the main optimal input parameter Ip = 8, voltage = 40, Ton = 200, duty cycle = 7 using copper tool during the machining of aluminium based hybrid metal matrix composite. The optimization of EDM parameter will save the time for machining and improve the surface roughness.

References

[1] M. Rezwee, (2019), "Electric Discharge Machining method for various Metal Matrix Composite materials", *International Journal of Innovative Technology and Exploring Engineering (*IJITEE). ISSN: **2278**-3075, Volume-8 Issue-9.

[2] M. Shukla,(2018) "Optimisation of Electrical Discharge Machining of Al-LM-6/ Sic/ B4C Composite: A Grey Relational Approach". *Materials Today: Proceedings* 5 **19147**–19155

[3] V. Suresh and A. Daniel, (2020) "Analysis of machining of parameter of aluminium alloy(LM25) graphite (Gr) boron carbide (B_4C) hybrid composite using wire EDM".*Materials Today: Proceedings.*

[4] M. Mastana and Smeutirekhapani, "Experimental analysis of AA6063aluminium alloy using EDM" *Materials Today: Proceeding*

[5] S. Manish and, S. Shukla and S.K Dhakad, (2018) "Optimization of electric discharge machining of AL-LM-/5/SiC/B_4C composite using GRA"*Materials Today: Proceedings* 5 **19147**–19155.

[6] Alip Kumar and Md. Yeasin, "Microstructural and machno-tribological behaviour of Al-6063 reinforcement SiC /TiC hybrid MMC", *Materials Today: Proceeding.*

[7] L. Natrayanand M. Senthil Kumar, (2019) "Optimization of wear behaviour on AA6061/Al2O3/SiC metal matrix composite using squeeze casting technique – Statistical analysis", *Materials Today: Proceedings,* pp. **2214**-7853.

[8] Arun Kumar Sharma, Rakesh Bhandari, Amit Aherwar and Ruta Rimasauskiene, (2019)"Matrix materials used in composites: A comprehensive study", *Materials Today: Proceedings*, pp. **2214**-7853, 2019.

[9] Ravindra Yadav, Ravi Kant Gupta and Ashish Goyal, (2020), "Study of tribological behaviour of hybrid metal matrix composites prepared by stir casting method", *Materials Today: Proceedings*, pp. **2214**-7853.

[10] Kamlesh, "A review paper on machining of metal matrix composite and optimization method used in electric discharge machining", *Materials Today: Proceedings* 5 (2018) 24428–**24438**. 2017.

[11] S. Marichmy and B. Stalin, (2020) "Optimization of machining parameters of EDM for α-β brass using response surface methodology". *Materials Today: Proceedings* **24** 1400–1409 2018.

[12]A.Adalarsan,(2015),"Optimization of laser cutting parameters for Al6061/SiCp/Al2O3composite using grey based response surface methodology (GRSM)", *Measurement:* **73** 596–606.

[13] A. Dey and Pandey. K. M, (2017) "Optimization of electrical discharge machining process parameters for Al6061/Cenosphere composite using grey-based hybrid approach". *Trans. Nonferrous Met. Soc. China* **27** 998–1010.

[14] M.B.A. Shuvho a , M.A. Chowdhury, (2020), "Surface characterization and mechanical behavior of aluminium based metal matrix composite reinforced with nano Al_2O_3,SiC,TiO_2 particle",*Chemical Data Collections* **28** 100442.

[15] S. Mandeep, "Investigating the EDM parameter effects on aluminium based metal matrix composite for high MRR". *Materials Today: Proceedings* xxx (xxxx) xxx.

[16] Mehrdad Shyan, Beitallah Eghbhali, (2020), "Fabrication of AA2024-TiO_2 nano composite through stir casting process", *Trans. Nonferrous Met. Soc. China* **30** 2891–2903.

[17] Bibin K Thariana, (2019) "Multi-Objective Parametric Optimization in EDM Using Grey Relational Analysis", *Materials Today: Proceedings,* **16** 1013–1019.

[18] K. Jitendra, (2019) "Multi-response optimization of machining parameter for Zircaloy by response surface methodology and grey relation analysis". *Materials Today: Proceedings xxx (xxxx) xxx.*

[19] M. Sivaraj, S. Rajkumar,(2020), "Parametric optimization of electric discharge machining Al-TiC composite using Grey relational analysis", *Materials Today: Proceedings* xxx (xxxx) xxx.

Energy Absorption of Glass Fiber Reinforced Plastic (GFRP)-Woven Composites in Different Nano Clay Loading Using Numerical Method.

[1*]Yared G. Zena, [1]Mulugeta H. Woldemariam

[1]Addis Ababa University, Addis Ababa Institute of Technology, School of Mechanical and Industrial Engineering, Ethiopia

Corresponding author mail: yared.getahun@aait.edu.et

Abstract: In this work, the numerical investigation for impact behavior of GFRP (Glass Fiber Reinforced Plastics) composite along modified with nano-clay (Cloisite® 20B) at 0.5%wt., 1%wt., and 2%wt concentrations was done. The finite element analysis was carried out in ABAQUS/Explicit to investigate the low-velocity impact response. The numerical results showed that adding 0.5% wt. and 1%wt. nano-clay significantly increased the stiffness of GFRP compared to a pristine one. In contrast, adding 2%wt nano-clay decreased the stiffness of GFRP. With regard of energy absorption, a 0.5%wt. nano-clay modified GFRP showed a maximum energy absorption, whereas, a 2 %wt. nano-clay modified GFRP exhibited the lowest performance for all impact energy levels.

Keywords: Glass fiber reinforced plastics (GFRP), Cloisite® 20B, nano clay, woven, impact

1. INTRODUCTION

For the last decades, the use of laminated composites especially glass fiber reinforced composites is highly used for improvement of advanced structure and cruel areas like oil walls is increased. And the need is also come to table to study low velocity impact behavior and damage. [1] [2]. To study the delamination of thick composite laminates under the low-energy impact, damage forecast and investigation were displayed in [3]. The simulation comes about appeared a great relationship to the experimental perceptions. The delamination area was restricted by the affected zone and the fiber dispersion.

A low-velocity impact test was done in carbon fiber/epoxy composite laminates reinforced by short fiber and other material to investigate damages in [4] and the results showed that the composite experience a various fracture, delamination, Intra ply cracking, matrix cracking, fiber breakage, and damage depending on the interlayer material. From experiment done in [5] different binder volume which is 3%wt. and 6%wt. have little effect in in-plane stiffness (twist and weft heading). delamination resistance for different matrix for preparing GFRP also was investigated by [6]. And conclude that for low velocity, the same impact condition polyurethane-based GFRP created a little damage measure and impact toughness, but two GFRPs appeared small contrast within the add up to energy absorbed during the impact.

Several researchers have proposed different failure theories to model the failure envelope of various types on composite materials among those failure criterion Hashin and puck failure criteria taken as good predictors of damage for laminated composites **[7] [8]**. Hence, in this work the constitutive intralaminar modes for studying the damage initiation and propagation of Cloisite® 20B nano clay modified GFRP were joined into the ABAQUS/Explicit FE code by user-defined VUMAT material subroutines to adopt the 3D Hashin criteria.

2. MATERIALS

As a material GFRP woven composite is used as a materials for structure crashworthiness analysis. The GFRP – plain weave composites are four types the first is a pristine material that is free of nano clay. The second is 0.5 %wt. per volume 20B nano clay modified, the third is 1%wt. per volume 20B Nano clay modified, and the fourth 2%wt. per volume 20B Nano clay modified. In this work, S-glass plain weave fabric and Cloisite® 20B (Southern Clay Products, Inc., TX) were used as a material and reported in [9] and mechanical properties [10], as shown in table 1 and table 2

Table 1. Tensile results of mechanical properties [10]

Direction	Warp				weft			
Loading	Pristine	0.5wt%	1 wt%	2 wt%	Pristine	0.5wt%	1 wt%	2 wt%
Tensile strength (MPa)	513.9	552.9	529.99	480.99	510.6	515.5	545.1	465.9
Tensile modulus of elasticity (GPa)	25.0	23.1	23.5	20.6	23.2	26.5	24.4	20.7
Tensile poisson's ratio	0.121	0.124	0.089	0.082	0.120	0.153	0.089	0.072

Table 2. Compressive results of mechanical material properties [10]

	Warp				weft			
Loading	Pristine	0.5wt%	1 wt%	2 wt%	Pristine	0.5wt%	1 wt%	2 wt%
Compressive strength (MPa)	86.3	77.2	83.5	97.7	92.7	77.1	76.9	91.7

Compressive modulus of elasticity (GPa)	25.2	25.9	26.3	28.3	24.8	25.1	26.1	22.2
Compressive poisson's ratio	0.058	0.086	0.060	0.078	0.101	0.091	0.065	0.079

3. MATERIAL MODELLING

The damage initiation and advancement show applied in this work were based on the Hashin 3D failure criteria. In this demonstration, four distinctive modes of failure were considered, to be specific, fiber crack in tension, fiber buckling and kinking in compression, matrix splitting beneath transverse tension and shearing, and matrix pulverizing beneath transverse compression and shear.

The equations used in this paper is based on Hashin 3D formulation and are taken from [3,8] and the short form are listed below

Where:

X^T *– tensile strength in longtudnal direction*
X^C *– compression strength*
S^L *– shear strength in longtudnal direction*
S^T *– transvers shear strength*
α is a coefficient of shear stress contribution to the fiber tensile failure initiation criterion; and
$\sigma_{11}, \sigma_{22}, \sigma_{33}, \sigma_{12}, \sigma_{13}$ and σ_{23} are components of the effective stress tensor, σ, that is used to evaluate the initiation criteria.

4. NUMERICAL CONFIGURATION

The drop-weight analysis was done based on ASTM D7136[11]. 12.7 mm distance across with side of the equator head impactor was utilized all through the quasi-static space and affect tests [12]. The mass of the impactor for this numerical analysis is 23.55 kg. based on Research Council for Automobile Repair (RCAR) low velocity consideration is also (5-20 km/h). The material properties were taken from [9] [10].
Finite element simulation of GFRP is based on ASTM – Designation D7136M-15 and the fixtures are shown in figure 1.

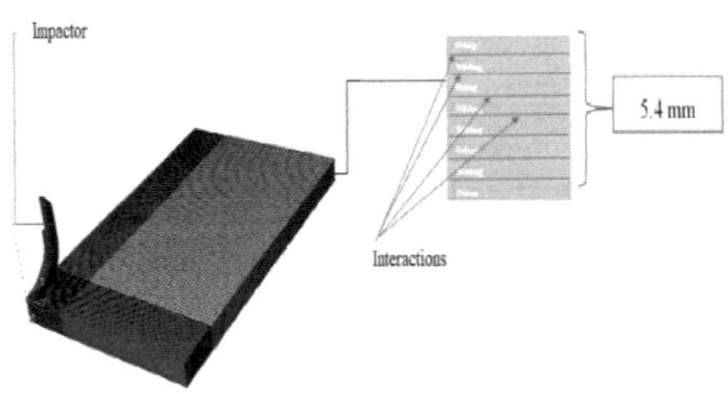

IOP Conf. Series: Materials Science and Engineering 1170 (2021) 012009 doi:10.1088/1757-899X/1170/1/012009

Figure 1. numerical specimen

The finite element simulation of all nano clay modification GFRP-WOVEN composites is based on ASTM - Designation D7136 M-15 with dimension 62.5 x 37.5 x 5.4 mm with 8 laminas. Each lamina is 0.675 mm thickness. The impactor diameter was 16 mm, the impact velocities used in this work were 1.2 m/sec, 1.7 m/sec, and 1.9 m/sec. To reduce the simulation time, a quarter of a GFRP specimen and impactor were used.

5. RESULT AND DISCUSSION

Figure 2 shows the comparison of reaction forces for pristine 0.5%wt. nano clay and shows 0.5%wt. nano clay loading shows a higher reaction force than pristine material. From the figure, the peak load for each material configuration reaches at approximately equal interval, which indicates that no significate damages were observed with 28.8 J impact energy. Figure 3 shows the comparison of pristine, 0.5%wt., 1%wt., and 2%wt. nano clays at 72.2 J impact energy. Results show that highest reaction force is gained from 0.5%wt. Nano clay modified and the second highest reaction force is pristine one., thirdly 1%wt Nano modified and the least reaction force is 2%wt. Nano clay modified GFRP. The 2%wt. nano clay loading reaches its maximum reaction force first and is followed by the pristine. This implies that the inclusion of excessive nanofillers for the present material, S-glass GFRP, has a notable effect on the damage resistance. On other hand 0.5 %wt. Nano clay inclusion increased the reaction force and improved the damage résistance.

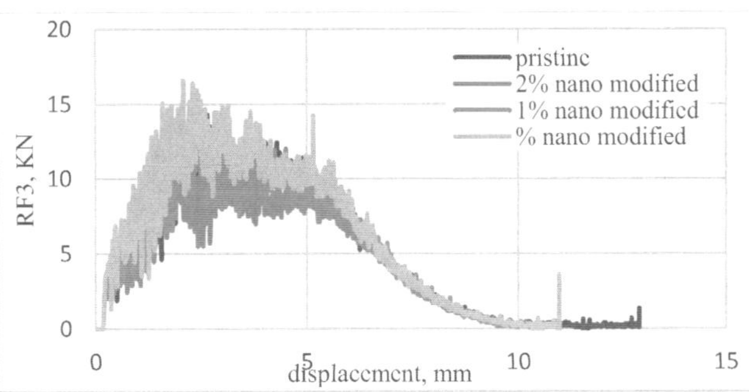

Figure 2. Comparison of 0%wt., 0.5%wt., 1%wt. and 2%wt. nano

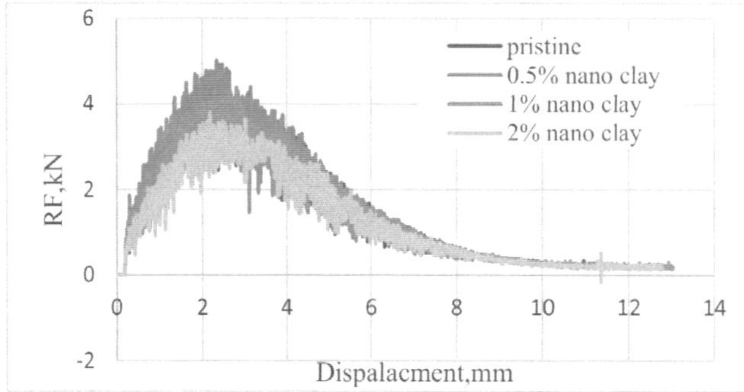

Figure 3. Comparison of pristine, 0.5%wt. ,1%wt. and 2%wt. nano clay at 28.8J

The affect damage resistance plots appeared in Figure 2, can be characterized by two limits, early stack, and crest stack. The beginning stack is recognized by the primary sudden drop within the reaction-time bend which shows the primary fabric harm such as interface disappointment or matrix splitting which is show ordinarily at/or close the back confront of laminates [14]. From figure 2 and figure 3 incipient load of 2%wt. nano clay is first followed by 1%wt. and 0.5%wt. nano clay loading and pristine is the last

As per [13] adding nano clay with 1%wt. loading increase, the energy absorption capacity, while [14] concludes samples with 1.5 wt.%wt. nano clay loading absorbs less energy as compared to 3 wt.%wt. and GFRE (0%wt.) samples. But in this research, the capacity of energy absorption in comparison with pristine is dependent on the amount of percent of nano clay. Based on Figures 4, 5, and 6, the energy absorption of 0.5%wt. nano clay exhibited a more energy absorption capability, with a variation of 5 to 10 %wt. for different impact energies, compared with the pristine, 1%wt., and 2%wt. nano clay modified GFRP.

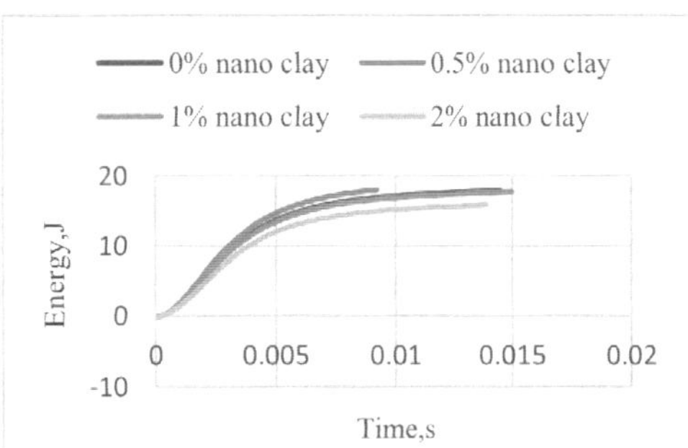

Figure 4. Energy vs time graph in 39.2J

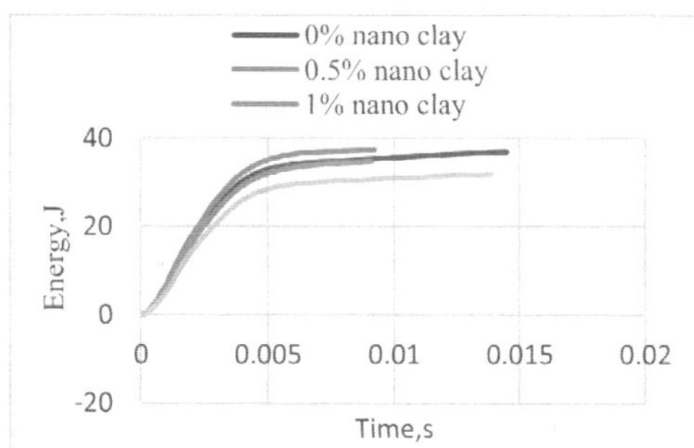

Figure 5. Energy vs time graph in 39.2J

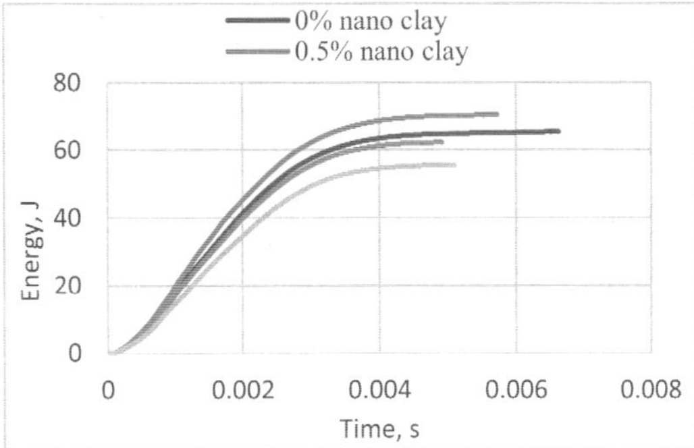

Figure 6. Energy vs time graph for 72.2 J

In [15], stiffness of the material is indicated by displacement or deflection at peak load which is the higher deflection the smallest in stiffness. As shown in Figure 7, the addition of nano clay with 0.5%wt. and 1%wt. nano clay increases the stiffness while adding nano clay with 2%wt. nano clay decreases the stiffness. As proven in [10], the results have a direct relationship to the modulus of elasticity. The modulus of elasticity in transverse direction for 0.5 and 1%wt. showed increased values, respectively, compared with the pristine one. On the other hand, 2%wt. nano clay showed the lowest transverse modulus of elasticity. As a result, the stiffness of 2%wt. nano clay reduced significantly.

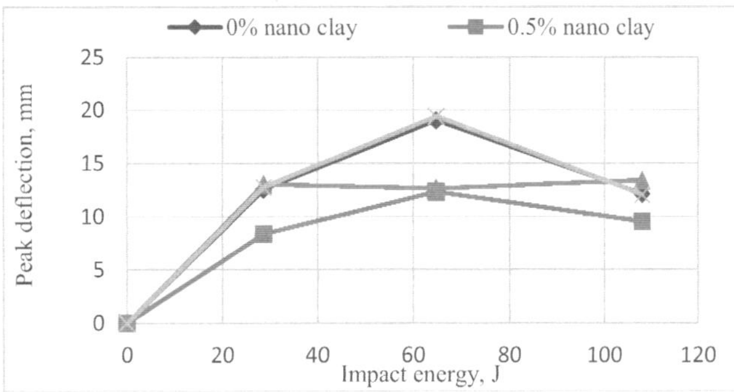

Figure 7. Variation of peak deflection with impact energy

6. CONCLUSION

The following conclusion is drawn from this research.

1. loading Cloisite® 20B nano clay with 0.5 %wt. shows good energy absorption than loading it 1%wt., 2%wt. and 0%wt. of GFRP.
2. The addition of nano clay with 0.5%wt. and 1%wt. nano clay increases the stiffness while adding nano clay with 2%wt. nano clay decreases the stiffness. Increased Young's modulus observed in stiff material and, in this work as validation 0.5%wt. nano clay has the highest

modulus of elasticity of all material in the transverse direction followed by 1%wt. nano clay loading. 2%wt. nano clay loading has a list in modulus of elasticity and stiffness.

3. Optimal concertation of nano clays introduces the desired phenomenon, increased energy absorption. A better and novel design is drawn from understanding of such situation in improvement of damage and impact behavior.

7. REFERENCE

[1] S. Eslami and F. Taheri, 2013 "Effects of perforation size on the response of perforated GFRP composites aged in acidic media," *Compos. Sci.*, **Vol. 69**, pp. 262–269, 201

[2] Sun, X. C., M. R. Wisnom, and S. R. Hallett. 2016. "Interaction of Inter- and Intralaminar Damage in Scaled Quasi-Static Indentation Tests: *Part 2 - Numerical Simulation." Composite Structures* **136**:727–42.

[3] O. Cousigné, D. Moncayo, D. Coutellier, P. Camanho, and H. Naceur, "Numerical modeling of nonlinearity, plasticity and damage in CFRP-woven composites for crash simulations," 2014.*Compos. Struct.*, **vol. 115**, no. 1, pp. 75–88,

[4] Sohn, M. S., X. Z. Hu, J. K. Kim, and L. Walker. 2000. "Impact Damage Characterisation of Carbon ® Bre / Epoxy Composites with Multi-Layer Reinforcement." *Carbon* 31(8):681–91.

[5] Gerlach, Robert, Clive R. Siviour, Jens Wiegand, and Nik Petrinic. 2012. "In-Plane and through-Thickness Properties, Failure Modes, Damage and Delamination in 3D Woven Carbon Fibre Composites Subjected to Impact Loading." *Composites Science and Technology***72(3)**:397–411.Retrieved

[6] Kuboki, T., P. Y. B. Jar, and T. W. Forest. 2003. "Influence of Interlaminar Fracture Toughness on Impact Resistance of Glass Fibre Reinforced Polymers." *Composites Science and Technology* **63(7):**943–53.

[7] Tsai, Steve and Ed Wu. 2009. "The Tsai-Wu Tsai Wu Failure Criterion The Tsai Wu Failure Criterion." (1939).

[8] Hashin, Z. and A. Rotem. 1973. "A Fatigue Failure Criterion for Fiber Reinforced Materials." *Journal of Composite Materials* 7(4):448–64.

[9] Koricho E.G., Haq M., Cloud G.L. (2017) Numerical and Experimental Characterization of Hybrid Fastening System in Composite Joints, **Volume 10.** *Conference Proceedings of the Society for Experimental Mechanics Series.* Springer, Cham.

[10] Mulugeta H. Woldemariam, 2020. "Influence of Cloisite® 20B on mechanical properties of S-glass fiber-reinforced polymer composite. *Proceedings of the Institution of Mechanical Engineers, Part C: Journal of Mechanical Engineering Science* **Vol 234, Issue 15**

[11] Method, Standard Test. 2002. "Standard Test Method for Tensile Properties of Polymer Matrix Composite Materials 1." **15.**

[12] Riccio, A., S. Saputo, A. Raimondo, and A. Sellitto. 2017. *Modeling Low Velocity Impact Phenomena on Composite Structures.* Elsevier Ltd.

[13] Koricho, Ermias G. et al. 2015. "Effect of Hybrid (Micro- and Nano-) Fillers on Impact Response of GFRP Composite." *Composite Structures* 134:789–98. Retrieved

[14] Rafiq, Ahmad, Nesar Merah, Rachid Boukhili, and Muneer Al-Qadhi. 2017. "Impact Resistance of Hybrid Glass Fiber Reinforced Epoxy/nanoclay Composite." *Polymer Testing* 57:1–11.

[15] Hosur, M. V., M. Adbullah, and S. Jeelani. 2005. "Studies on the Low-Velocity Impact Response of Woven Hybrid Composites." *Composite Structures* 67(3 SPEC.ISS.):253–6

Investigation of Thermal Response of Disc Brake System: A Review

Sachin Thakre[1], Dr. Achal Shahare[1] and Dr. G. K. Awari[2]
[1] Mechanical Engineering Department, G H Raisoni University, Amravati, India
[2] HOD, Automobile Engineering, Government Polytechnic, Nagpur, Nagpur, India
E-mail: thakresachin2k1@rediffmail.com

Abstract. A comprehensive review of the thermal response of disc brake system has been made. Extensive research has been carried out on the operating problems, the material for the rotor, design of the rotor and the braking conditions of the disc brake systems and numerous methods of improvement are proposed. Various ventilation patterns havebeen studied in order to enhance thermal efficiency, and structural performance reductions have been found in most cases. Which creates need for the design of discs with optimal thermal and structural characteristics, so that disc brake performance can be enhanced without compromising in strength and other parameters of the disc brake system. It is expected that this exhaustive and comprehensive review will help the research fraternity in identifying the research gaps for their future work.The prime purpose of this paper is to provide an exhaustive overview of developments going on in this field. The proposed research work attempts to develop and modify the system whichuses the effective method to minimise the temperature and weight of disc brake rotor bythermo-mechanical analysis to optimise the performance of Disc brake. The proposed design of disc brake offers a promising solution to improve the braking efficiency by 15%and overall reduction in weight of disc brake by 10%.

Keywords. Disc Brake, Thermal Response, Rotor, Finite Element analysis.

1. Introduction

The braking system is an important safety and control systems of an automobile. It is used to reduce the speed of the vehicle from an actual speed to a desired. It also used to stop or hold the vehicles. [1] Because of many advantages of disc brakes over drum brakes have led to their use in passenger-cars and light-trucks. Most automotive are fitted with disc brake systems. Two types of disc brakes are employed in various makes of disc brakes i.e. the solid and ventilated types. The calliper is attached to the axle casing of the vehicle. When the brakes are applied friction pads move towards the disc applying equal and opposite forces on disc. Friction pads convert the kinetic energy of the vehicle into heat energy through friction. In the present research work, two friction pads are used in the brake with their friction surfacesfacing the rotor. There are four main types of brake pads used in vehicles: non-asbestos organic, ceramic and semi-metallic and metallic. Out of these metallic brake pads are used in most of the vehicles. They are typically made of iron, copper, steel and graphite all mixed and together and bonded to form the pad material. [4, 44,]

The efficiency of the braking system is influenced by the temperature increase of the disc and the friction pad interface during braking. [13] The temperature rise causes thermal distortion, brake fading, brake fluid vaporisation and brake squealing, etc. [16] All of these factors straight away affect the life and performance of the brake system and, most importantly, the safety of the vehicle and the passenger. [41, 46, 51]
The literature review reveals the importance of the thermo mechanical analysis of disc brake and

IOP Conf. Series: Materials Science and Engineering 1170 (2021) 012010 doi:10.1088/1757-899X/1170/1/012010

to reduce the rotor weight of disc brake regarding the optimisation of disc brake performance. The following areas are worth focusing

- To optimize design and ventilation patterns of brake discs.
- Materials and thermo-mechanical analysis along with structural dynamics can be studiedby considering air cooling effect on brake discs. [41, 26]

In section 2 the basic information about disc brake is explained along with its components and requirements. Section 3 consolidates the research works on thermal and mechanical behaviour of disc brake. In sections 4 performance analysis of disc brake on the basis of design and types of disc brake rotors is presented. In sections 5 performance analysis of disc brake on the basis of operating problems and braking criteria of disc brake is presented. In sections 6 performance analysis of disc brake on the basis of thermo-mechanical analysis of disc brake is presented. In sections 7 performance analysis of disc brake on the basis of materials of disc brake rotor is presented. In last sections challenges and future direction is covered.

1.1. Objectives

The prime objectives of this research paper are as follows:

- To identify the research gaps for further research.
- To design and develop the disc brake system for two wheelers using thermal analysissystem approach.
- To investigate the performance of disc brake by opting the various geometry of rotor.
- To optimize the performance of disc brake.

2. Brakes used in Vehicles

Brake is a device that transforms the kinetic energy of the running automobile into heat by means of friction. In braking, the friction pad and disc takes away the speed from wheels. The energy consumed by the brake is the thermal energy and the heat is dissipated into the atmosphere and the vehicle's motion is stopped. [2, 18]

Brakes are the important control systems of the vehicle. It controls the vehicle in the shortest possible time and distance. It also control the speed of vehicles at turning and other crowded places.

2.1 Materials of Disc Brake

Disc brake rotors are made from gray cast iron, contains 3.5% carbon and other additives. [23] Carbon fibre reinforced carbon–silicon carbide (Cf/C–SiC), a ceramic composite, has been used as a substitute for grey cast iron. Metallic ingredients are added to improve wear resistance, heat transfer and mechanical strength. [22, 14]

2.2 Constructional features of Disc brake

The disc brake as shown in Fig. 1 is a brake that uses callipers to press the pads against the disc to produce friction that reduces the rotation of the wheels, either to minimise their speed or to keep them stationary. Disk brake elements are as follows Disc or Rotor, Caliper, Brake pads, Master cylinder, Brake fluid reservoir, Brake line and Brake fluid. [9, 28]

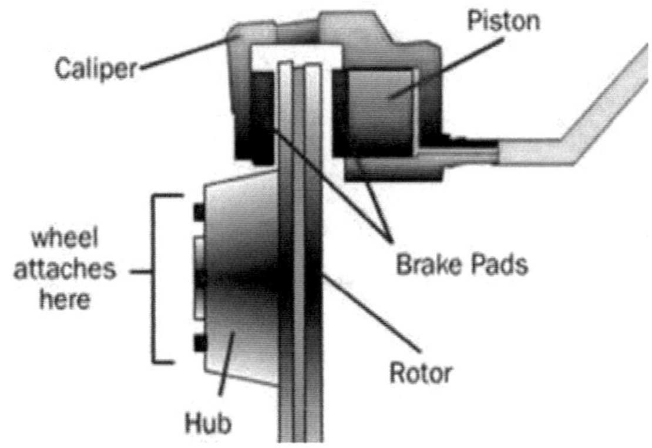

Fig. 1 - Disc Brake [11]

2.3 Disc brake Rotors

It is the component which is used to hold or to stop the wheel rotation. Rotors of disc brake is made of cast iron. In racing cars and high performance vehicles composite materials such as carbon fibre or ceramics rotors are used. Fig. 2 shows the assembly of disc brake which includes Rotor, Caliper and two brake pads and it explains how these parts are interconnected. [5, 9]

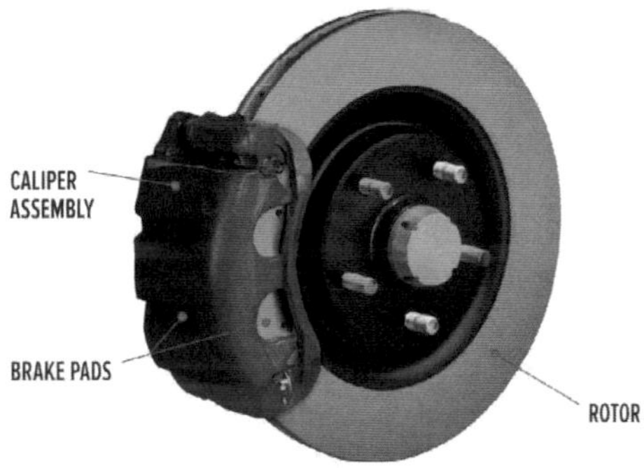

Fig. 2 – Rotors and brake pad assembly

3. Thermal stress analysis of Disc Brake

In order to study the thermal analysis of the brake discs under the conditions of cyclic thermal load, the temperature distribution of these brake discs is a function of time. [34] The heat produced at the disc-pad interface during sudden braking allows for non-uniform distributionsof temperature in the rotor. The heat energy released varies from 300 to 800 C. [4]

The output performance of the vehicle with respect to the braking ability depends on the friction coefficient at the brake pad and disc interface. As temperature increases, the brake disccauses the friction coefficient of the brake pads to increase and improves the braking torque output. Whereas the elevated temperatures have an impact on disc and pad wear. [47, 49, 50] The friction coefficient of the brake pads unexpectedly decreases at elevated temperatures (approximately 550 ºC), resulting in a reduced braking torque output. The risk of thermal cracking is growing at these temperatures. It is therefore important to keep the brake surface temperatures within acceptable limits (i.e. below 500 ºC), in order to ensure the efficient operation of the vehicle braking system. [19, 20, 33]

Daanvir Karan Dhir [4], proposes that geometry of disc is an important factor in evaluating its thermal and mechanical properties. Author indicates the use of discs with holes and air foil vents where faster cooling of braking system is required, like in the racing cars and bikes. [4]

3.1 Heat Dissipation

The various modes of heat transfer through which heat is dissipated such as conduction, convection and radiation are explained in Fig. 3.

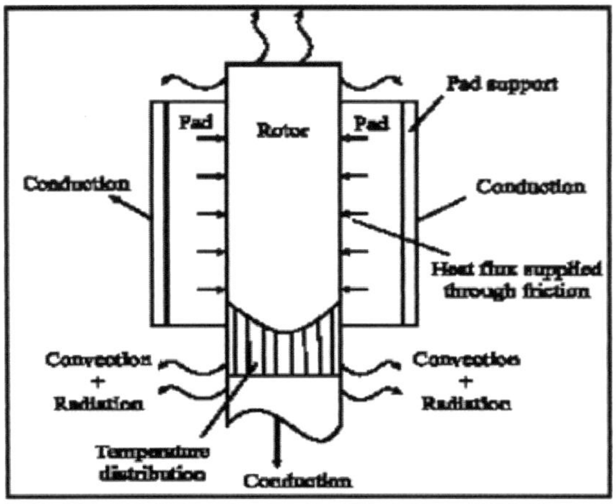

Fig. 3 Heat transfer in disc rotor [4, 44]

3.2 Transient Thermal Analysis

Transient thermal analysis can be used to find out the temperature and other related quantities that vary over time. Changes in temperature distribution are required in many applications, such as the quenching analysis for heat treatment. The elevated temperature distribution results in increased thermal stress, which may be the explanation for the failure of such systems. The transient thermal analysis for thermal stress assessment plays a very important role in such situations. [2, 33, 36]

Deepak S. Hugar et al. [2], reported useful design and help to improve the thermal, mechanical and structural performance of disc brake system.

Table 1 shows the comparison of deformation and stress for various discs.

Table 1. Comparison of deformation and stress [2]

Disc	Deformation	Stress
Original Disc	0.00369	19.08
New disc 1	0.00382	19.67
New disc 2	0.00568	29.91
New disc 3	0.00534	27.45
New disc 4	0.00351	18.00
New disc 5	0.00585	25.65

Table 2 shows the comparison of weight and temperature for various discs.

Table 2. Comparison of weight and temperature [2]

Disc	Weight(Kg)	Temperature(°c)
Original Disc	0.98	124.93
New disc 1	1.08	123.03
New disc 2	0.87	231.98
New disc 3	0.96	165.72
New disc 4	1.14	128.20
New disc 5	0.89	95.60

S.A.M Da Silva et al. [5], have reviewed the comparative analysis of the grooved disc brake rotor and the drilled-grooved disc rotor and the author's results are as follows. The holes and grooves are provided for a cooling method and to have a greater grip on road during braking. The results obtained by authors regarding the criteria, various point load to both brake discs as shown in Table 3 and Table 4.

Table 3. Displacement and stress for a drilled and grooved disc. [5]

Drilled and grooved brake disc	Load(KN)	Displacement (m)	Von Mises Stress (MPa)
Column A	20	0.00001938	57.249
Column B	23	0.00002306	65.982
Column C	26	0.00002694	74.423
Column D	29	0.00002881	83.011
Column E	32	0.00003269	91.599
Column F	35	0.00003557	100.18

Table 4. Stress and displacement for a grooved disc. [5]

Grooved brake disc	Load(KN)	Displacement(m)	Von Mises Stress (MPa)
Column A	20	0.0000236	69.26
Column B	23	0.0000272	86.23
Column C	26	0.0000307	99.99
Column D	29	0.0000342	113.77
Column E	32	0.0000378	125.54
Column F	35	0.0000413	137.31

4. Performance analysis on the basis of design and types of disc brake rotor

A comprehensive review of the thermal response of disc brake system has been made and findings of this review and the performance analysis on the basis of design and types of disc brake rotor of disc brake is presented in this section. It also discusses the limitations about the studied research papers.

In this paper author carried out the study for different hole diameters of brake disc andobserved the deformation and stress developed. The deformation in the rotor plate of disc brake with elliptical holes is 8.4% higher than that of the stress developed in the rotor plate of disc brake having equivalent circular holes. The stress developed in the elliptical holes rotor plate of disc brake is 2.5% less than that of the rotor having equivalent circular holes. However, the shape of hole is changed from elliptical shape to equivalent circular shape and analyzed its effect only on two wheeler disc brake. More parameters could take into considerations for better results. [12]

In this paper author carried out the study on two wheeler Bajaj Discover 220cc and the following parameters Weight, Deflection, Shear Stress and Factor of Safety is observed. The author suggests new disc weight is 26.68 % lesser than the original disc. However, only mathematical approach is taken into considerations. [3]

This paper conducted study on various types of discs on the basis of geometric design. The author suggests that disc rotor 1 has weight 15.18 % higher than the rotor and a temperature rise of 12.32 percent higher than the disc 3 temperature. However, this study suggests the use of disc flange rotors only for restricted areas of heavy braking where a higher braking force is necessary. [4]

This paper conducted study on various types of discs such as Grooved and drilled-grooved disc rotor and the parameters displacement and stress were observed. Author concludes that displacement experienced by the brake disc is directly proportional to the stress on a sections. However, only application of linear load is taken into considerations for analysis. [5]

This paper conducted study on various types of discs such as full and ventilated type brake discs and the parameter temperature distribution is observed. The author concludes that the radial ventilation plays significant role in the cooling of brake discs during braking. However, validations through experimental study is not done by author to check the accuracy of the numerical model developed. [43]

5. Performance analysis on the basis of operating problems and braking criterion

A comprehensive review of the thermal response of disc brake system has been made and findings of this review and the performance analysis on the basis of operating problem and braking criteria of disc brake is presented in this section. It also discusses the limitations about the studied research papers.

In this paper author carried out the study on continuous braking criteria and the following parameters Von Mises stress distribution and contact pressure distribution were observed. The author suggests that the ventilated disc brake type is the best braking device for the specified application. The experimental values get from this analysis are below the limit of their allowable values. However, author suggests thermal-structural and contact analysis of a brake model without its thermal analysis. [42]

In this paper author carried out the study on repeated braking criteria and the following parameters temperature, operating speed and coefficient of friction were observed. The author claims that compared to the simulation, the laboratory experimental investigation used for the predicted field results is in equal agreement. However, experimental analysis is carried out only for brake systems of two wheeler and without variations in speed of vehicles. [27]

In this paper author carried out the study on panic braking criteria and the following parameters total deformation, stress and strain developed were observed. The author suggests that, structural analysis is carried on the caliper when the pressure will increase and the deformation is observed. Thermal analysis is done on the periphery of the pad and the convection film coefficient was applied to the overall surface of the disc as convection occurson the overall surface. However, Study is carried out only for existing braking component, the author does not focused on the parameters which needs to be considered for designs of new braking components except the panic braking criterion. [17, 25]

6. Performance analysis on the basis of thermo-mechanical behavior

A comprehensive review of the thermal response of disc brake system has been made and findings of this review and the performance analysis on the basis of thermo-mechanical analysis of disc brake is presented in this section. It also discusses the limitations about the studied research papers.

In this paper author carried out the study on numerical analysis of solid and ventilated discs and the following parameters temperature distribution and Von Mises stress distribution were observed. The author suggests that study has been carried out for various ventilation patterns to improve the thermal performance, but decrease in structural performance was found in most of the cases. Which creates the need for new research in this area. However, researchers have studied different parameters according to the objectives of the analysis. The question remains, however, which key factors need to be taken into account for a practical and thoroughstudy of the disc brake. [41]

In this paper author carried out the study on structural and thermal analysis of the disc brake and the following parameters vehicle geometries, materials and the dimensions of the brake components were observed. The author reveals that vaporization of brake fluid can cause due to an increase in the temperature of the pad and therefore increase the ability of the brake fluid to heat up. However, validations through experimental study to check the accuracy of the analytical model is not done by the author. [35, 40]

This paper conducted study on structural and thermal analysis of the disc brake and the following parameters equivalent stress, temperature distribution and total deformation were observed. Author concludes that the cast iron gives better performance than other materials regarding stress point of view however stainless steel gives better brake performance than others from deformation point of view. However, author carried out study only for threematerials. [48]

This paper conducted study on structural and thermal analysis of the disc brake and the following parameters heat dissipation and strength were observed. The author concludes that heat dissipation is better for the aluminium but as the strength of aluminium is poor, hence this material cannot be used for the disc brake rotors. However, author carried out study only for three materials. [52]

This paper conducted study on thermo-mechanical analysis of the disc brake and the following parameters temperature, stress, total deformation and contact pressure distribution were observed. The author claims that at the outer radius of brake disc, more deformation occurs. However, many thermal and mechanical factors would have been taken into considerations to interpret its impact on fatigue life of brake discs. [45]

7. Performance analysis on the basis of materials of disc brake rotor

A comprehensive review of the thermal response of disc brake system has been made and findings of this review and the performance analysis on the basis of materials of disc brake rotor of disc brake is presented in this section. It also discusses the limitations about the studied research papers.

In this paper author carried out the study on two wheeler Bajaj Pulsar 220cc and the following parameters deformation, temperature and weight is observed. The author suggests that grey

cast iron is the best disc brake material. However, only one type of rotor material i.e. gray cast iron is taken into considerations for studying on Bajaj Pulsar Vehicle. [2]
In this paper author carried out the study on ceramic material of disc brake rotor and the following parameters tribological properties and strength/weight ratio is observed. Author proves Al2O3 have better thermal and tribological properties for the said applications. However, author claims that, joining of Al2O3 to aluminium substrate presents major challenges as the thermal stresses are more. [6, 10]
In this paper author carried out the study on laser surface nitride Ti–6Al–4V material of disc brake rotor and the following parameters mechanical and tribological properties were observed. The author suggests that, tribological characteristics of laser nitrided Ti-alloy shows good performance, allows the possibility of using it as a light weight rotor than the heavier materials. However, several mechanical and tribological properties could take in to considerations to have better results and to minimize environmental and health issues. [08, 15,21]
In this paper author carried out the study on light weight AMC brake rotor material of disc brake rotor and the following parameters energy consumption, weight and cost were observed. The author claims that, for a fixed geometry of brake rotor, the application of light weight AMC material gave a 50% weight reduction. For this amount of weight reduction, the predicted energy savings is 19% for the considered brake rotor. However, author carried out study only for two materials. [07]

8. Challenges and research directions

It is inferred from the said analysis that a variety of factors are responsible for the efficiency of the disc brake. Researchers have tested different criteria according to their research objectives. The issue remains as to the primary factors should be taken into account for a practical and thorough study of the disc brake. [41] From a literature study, it is found that some of the researchers have already carried out the task of reducing the weight of the rotor, different possibilities of geometric variations of the rotor, adjustment of different rotor materials to improve the efficiency of the disc brake. It indicates, however, that there are some lacunas:

- A comprehensive investigation is required for effect of thermo mechanical stressconcentration and displacement on rotor of Disc brake.
- Further investigations are needed for optimisation of disc brake in Electric two wheelers.[29, 30, 32]
- Less attention has been paid to the brake fade issue of disc brake and effect of water film (hydrodynamic effect) on disc brake performance. [24]
- An elaborate investigation about the various geometric variations of disc brake isrequired.
- Extensive research is done on the concept of Brake Squeal but the problem is not completely solved. Need more research to be carried out in this area? [31,37,38,39]
- Hydrodynamic effect on disc brake performance has not been investigated. [24]
- Less study related to computational investigation and validation of rotor and calliper ofdisc brake is carried out. Further investigations are required to optimise the performance of disc brake.

The literature review reveals the importance of the thermo mechanical analysis of disc brake and to reduce the rotor weight of disc brake regarding the optimisation of disc brake performance. To overcome the above lacunas, the following areas are worth focusing:

- To optimize design and ventilation patterns of brake discs.
- Materials and thermo-mechanical analysis along with structural dynamics can be studied by considering air cooling effect on brake discs. [41]
  The propose research work attempts to develop and modify the system which uses the effective method to minimise the temperature and weight of disc brake rotor by thermo - mechanical analysis of disc brakes to optimise the performance of Disc brake.

9. Conclusions

A comprehensive review of the thermal response of disc brake system has been made. Care has been taken to review most of the available materials. Authors have extensively reviewed literature and provided directions for future work. The literature review reveals the importance of the thermo mechanical analysis of disc brake and to reduce the rotor weight of disc brake regarding the optimisation of disc brake performance.
Researchers have mostly used finite element analysis for the analysis of disc brake thermal, mechanical and structural behaviour. Various ventilation patterns has taken into consideration for improving thermal performance, but found decline in structural performance in most of the cases. Which creates need for the design of discs with optimal thermal and structural characteristics, so that disc brake performance can be enhanced without compromising in strength and other parameters of the disc brake system. The proposed design of disc brake offers a promising solution to improve the braking efficiency by 15% and overall reduction in weight of disc brake by 10%. Various studies have been reported on material transition, braking conditions, period of braking, assumptions, field of application, etc. However, this field of study is wide open for further improvement. It is anticipated that this thorough and comprehensive analysis will help the research fraternity in identifying the research gaps for their further work, and that it will also help automotive designers for designing better disc brake systems with optimum efficiency for the benefit of society.

10. References

[1] P Shiva Shanker, 2017, A review on properties of conventional and metal matrix composite materials in manufacturing of disc brake, ICMPC, 5 (2018) pp 5864–5869.

[2] Deepak S. Hugar, Prof. U. B. Kadabad, 2017, Design and Thermal Analysis of Disc Brake for Minimizing Temperature, (IRJET), Volume: **04**, pp 2395 -0056.

[3] Swapnil Thigale, Chinmay Shah, 2016, Weight Reduction In Brake Disc Using Topology Optimization, Volume: **05** Issue: 10, pp: 2321-7308.

[4] Daanvir Karan Dhir, 2018, Thermo-mechanical performance of automotive disc brakes, Proceedings5, pp 1864–1871.

[5] S.A.M Da Silva, DVV Kallon, 2019, FEA on different disc brake rotors, SMPM 2019, Procedia Manufacturing 35 181–186.

[6] M.A. Maleque, S.Dyuti and M.M. Rahman, 2010, Material Selection Method in Design of Automotive Brake Disc, Vol **III**, WCE 2010.

[7] M.A. Malequea, et al. 2012, Energy and Cost Analysis of Weight Reduction using Composite BrakeRotor, Int. J. Vehicle Structures & Systems, 4(2), pp 69-73.

[8] Muthukannan Duraiselvama, et al. 2014, Laser surface nitrided Ti-6Al-4V for light weight automobile disk brake rotor application, Wear 309 (2014) pp 269–274.

[9] Asim Rashid, 2014, Overview of disc brakes and related phenomena - a review , Vol. **10**, No. 4,257.

[10] L. Gorjan, M. Boretius, 2016, Ceramic protection plates brazed to aluminum brake discs

Ceramics,CERI13260.

[11] Hanamant Yaragudri & Rajesh A, 2017, Design and Stress Analysis of Disc Brake, IJA-ERA.

[12] G. Maruthi Prasad Yadav, and M Venkateswara Reddy, 2017, Modelling and Analysis of Two Wheeler Disk Brake for Optimum Performance, NCETME -2017.

[13] Matthieu Hascoiet, Loïc Adamczak, 2020, At source brake dust collection system, Elsevier Ltd. Results in Engineering 5 (2020) 100083.

[14] Ferdinand H. Farwickzum Hagen, 2019, On-Road Vehicle Measurements of Brake Wear Particle Emissions, Elsevier Ltd. AEA 116943.

[15] S.P. Jadhav, S.H. Sawant, 2019, A review paper: Development of novel friction material for vehicle brake pad application to minimize environmental and health issues, Elsevier Ltd, 2214-7853.

[16] Pavol Kurčíka and Juraj Gerlicia, 2019, Innovative solution for test equipment for the experimental investigation of friction properties of brake components of brake systems, TRANSCOM 2019, Transportation Research Procedia 40 (2019) 759–766.

[17] Gabriele Rivaa et al., 2019, An FEA approach to simulate disc brake wear and airborne particle emissions, Elsevier Ltd. 138 (2019) pp 90-98.

[18] Kateryna Kravchenko et al., 2019, Estimating the brake disc air resistance by scaled test rig and by computer simulation, TRANSCOM 2019, Transportation Research Procedia 40 (2019) 737–744

[19] T.J. Mackin, S.C. Noe et al., 2002, Thermal cracking in disc brakes. Engineering failure analysis, Permagon, 9(63-76).

[20] T. Okamura, H. Yumoto, 2001, Fundamental study on thermal behaviour of brake discs. SAE Paper 2006-01-3203, 2001.

[21] Bushra Rashid, 2017, Eco-Friendly Composites for Brake Pads from Agro Waste: A Review, Materials Science and Materials engineering doi:10.1016/B978-0-12-803581-8.10159-6.

[22] Guangyu Bian et al., 2015, Friction performance of carbon/silicon carbide ceramic composite brakesin ambient air and water spray environment, Elsevier Ltd, TribologyInternational92(2015)1–11.

[23] Peter J. Blau et al., 2007, Tribological investigation of titanium-based materials for brakes, Elsevier Ltd., Wear 263 (2007) 1202–1211.

[24] P.J. Blau et al., 2003, Effects of water films and sliding speed on the frictional behaviour of truckdisc brake materials, Elsevier Ltd., Tribology International 36 (2003) 709–715.

[25] Dr. A. Satyanarayana Reddy and Shaik Mudassir Mohiddin, 2017, Design and Analysis of DiscBrake and Calliper of Four Wheeler, IJSTE, Volume **3**, Issue 08, pp. 186-190.

[26] Dr. G. K. Awari et al., 2020, Design and Analysis of Two-Wheeler Disc Brake to Reduce Weight for Optimum Performance: A Review, ICGESD-2020.

[27] Dr. Ranjit Ganaptrao Todkar, et al, 2018, Design and Development of a Disc Brake Test Rig for Experimental Investigation of a Disc Brake Temperature under Repeated Braking Cycles, IJETSR, Volume **5**, Issue 04, pp. 206-213.

[28] Gosai Dipak C et al, 2016, Use of Disc Brake in E-Car: A Review, IJSTE, Volume **3**, Issue 05, pp. 185-190.

[29] Khaled Itani, et al., 2016, Comparison between two braking control methods integrating energy recovery for a two-wheel front driven electric vehicle, Elsevier Ltd., Energy Conversion and Management 122 (2016) 330–343.

[30] Joseph Godfrey, V. Sankaranarayanan, 2016, A new electric braking system with energy regeneration for a BLDC motor driven electric vehicle, Elsevier Ltd.,

[31] N.M. Kinkaid et al, 2003, Review Automotive disc brake squeal, Elsevier Ltd. Journal of Sound and Vibration 267 (2003) 105–166.

[32] Soniya.K.Malode, R.H.Adware, 2016, Regenerative Braking System in Electric Vehicles, IRJET, Volume **3,** Issue 03, pp. 394-400.

[33] Timothy Mew, 2014, Transient Thermal Response Of Solid, Pinned And Highly Porous

Ventilated Brake Discs, University of the Witwatersrand, Johannesburg.

[34] Saiprasit Koctniyom, 2000, Thermal Stress Analysis of Automotive Disc Brakes, The University of Leeds School of Mechanical Engineering.

[35] Shaun Zedi, 2018, The Thermal Response Of Brake Discs At Varying Speeds, University of the Witwatersrand, Johannesburg.

[36] Naga Vamsi Krishna Kora and Srivatsan Thuppal Vedanta, 2016, Heat Transient Transfer Analysis of Brake Disc/Pad System, Blekinge Institute of Technology Karlskrona, Sweden.

[37] Cheol Kim et al., 2018, Analysis of Low-Frequency Squeal in Automotive Disc Brake by Optimizing Groove and Calliper Shapes, International Journal of Precision Engineering And ManufacturingVol. **19**, No. 4, pp. 505-512.

[38] Takashi Nakae, et al, 2018, Effect of adding mass to rotor on in-plane squeal in automotive disc brake, MATEC Web of Conferences 211, 13006 (2018).

[39] C. KIM and K. ZHOU, 2016, Analysis Of Automotive Disc Brake Squeal Considering Damping And Design Modifications For Pads And A Disc, International Journal of Automotive Technology, Vol. **17**, No. 2, pp. 213–223.

[40] Faramarz Talati and Salman Jalalifar, 2009, Analysis of heat conduction in a disk brake system, Heat Mass Transfer, Springer.

[41] Asif Afzal et al., 2018, Thermo-Mechanical and Structural Performances of Automobile Disc Brakes:A Review of Numerical and Experimental Studies, Springer, https://doi.org/10.1007/s11831-018- 9279-y.

[42] Ali Belhocine, et al., 2016, Thermal and structural analysis of disc brake assembly during single stop braking event, Australian Journal of Mechanical Engineering, VOL . **14**, NO . 1, 26–38.

[43] Arun kumar et al., 2016, A Study of Modes of Heat Transfer on Disc Brake, IJARIIE-Volume **2**, Issue 02, pp. 1637-1642.

[44] Faramarz Talati and Salman Jalalifar, 2008, Investigation of heat transfer phenomena in a ventilated disc brake rotor with straight radial rounded vanes, Journal of Applied sciences.

[45] Ali Belhocine and Asif Afzal, 2020, Computational finite element analysis of brake disc rotors employing different materials, Australian Journal of Mechanical Engineering, ISSN: 1448-4846, 2204-2253.

[46] Kao T, Richmond JW, Moore MW, 1994, The application of predictive techniques to study thermo- elastic instability of brakes. In: 12th annual brake colloquium and engineering display. SAE Technical Papers 942087.

[47] Anderson AE, Knapp RA, 1990, Hot spotting in automotive friction systems. Wear 135(2):319–337.

[48] Viraj Parab, Kunal Naik, Prof A. D. Dhale, Structural and Thermal Analysis of Brake Disc, 2014, Published by IJEDR.

[49] Panier S, et al., 2004, An experimental investigation of hot spots in railway disc brakes. Wear256(7–8):764–773.

[50] Majcherczak D, et al., 2006, Dynamic analysis of a disc brake under frictional and thermomechanical internal loading. Arch Appl Mech 75(8–9):497–512.

[51] Jung SP, et al., 2011, Thermo-mechanical finite element analysis of hot judder phenomenon of aventilated disc brake system. Int J Precis Eng Manuf 12(5):821–828.

[52] Swapnil.D.Kulkarni, J.J.Salunke, 2017, Thermal Analysis of Brake Disc, IRJET, Volume **6**, Issue02, pp. 126-131.

IOP Conf. Series: Materials Science and Engineering 1170 (2021) 012011 doi:10.1088/1757-899X/1170/1/012011

Review on Reactivity Controlled Compression Ignition Engines: an Approach for BSVI emission Norms

V S Kumbhar[1], Dr. Achal Shahare[2], Dr. G K Awari[3]
[1,2] Department of Mechanical Engineering G H Raisoni University Amaravati, India
[3]Department of Automobile Engineering, Government

Polytechnic, Nagpur, IndiaE-mail: vijayskumbar@hotmail.com

Abstract. In BSVI diesel engines, the limits of the NOx (oxides of nitrogen) and PM (particulate matter) reduced by 68% and 82% as compared to BSIV engines for the category of the vehicle having gross weight less than 3500kg. It is subjected to implement a complex and costly emission reduction system which reduces fuel economy. Reactivity controlled compression ignition (RCCI) is a duel fuel combustion strategy which has great potential to reduce NOx and PM and the need for an advanced after treatment system with enhanced thermal efficiency. The paper reviews potential of the RCCI strategy, to achieve the emission standards of BSVI norms, which reflects the need for cost assessment of existing engines equipped with advanced after treatment technologies.

1. Introduction

Stricter emission norms and increasing fuel prices promote the demand of cleaner and efficientcombustion. As a part of these, current engine research focused on the new modes of combustion, alternative fuels, and blending and composition, new strategies of fuel supply system, and optimization of the current technology for the enhancement of fuel economy and reduction in exhaust emission.[5, 8,11,32,43,44,48,56]

Diesel engines have wider applications in the field of transportation as well as the energy because of nearly zero throttled charging, almost fuel-lean combustion, and utilization of highercompression ratio, high power density, etc.[2,8,45] Generally, in modern CI engines the diesel isinjected just before the TDC during a compression stroke.[11,28] The diesel fuel having a moderate cetane number as well as it is high viscous and low volatility, it is difficult to get higher degree of hominization of air-fuel mixtures before the combustion initialization. The combustion of diesel engines is consists of pre-mixed and mixing regulated combustion phases.[19,53]At the initial stage, when the mixture prepared during the ignition delay, there isa higher HRR (heat release rate) which nearly burns at chemically correct conditions. Result ofthis there is a sudden increase in both the pressure as well as temperature inside the cylinder enhances NOx emissions. In case of mixing controlled combustion process, the burning rate of the fuel is managed by the air-fuel mixing rate.[15,41] Due to the injection delay and the various properties of the diesel fuel, there is generation of the heterogeneous mixture which leads to forming fuel rich zones, responsible for the PM formation.[28,52] There are various adverse effects of the PM and NOx emissions as human health and the environment are concerned, like cardio diseases, acid rain, smog, greenhouse gas effect etc.[14,47]As a trade-off relation between the NOx-PM , in the case of conventional diesel engine, it is very difficult to achieve such a low emission levels.[19,21,51,57,64] There are various methodologies adopted to achieving the targeted limits as per the various regulations such as fuel modifications, use of alternative fuels, engine level changes and the exhaust after-treatment level changes. [17,43,60]As system integration and complexities in operations like reductant storage, passive regeneration, required near about 10% of fuel cost develops the burden on the engine manufacturer as well as the owner of the vehicle.[58] The RCCI technology reduces the need of complex and costly after-treatment system utilized for the reduction of the NOx and PM.[35]

1.1. *Objectives*

The objectives of this paper are as follows

1.1.1 Investigation of limitations of the conventional diesel engines as emissions is concernedand historical aspects of RCCI.

1.1.2 Study of the chronological development of RCCI engine.

1.1.3 Persual of advanced emission reduction systems and their critical aspects.

1.1.4 Detailed review of critical parameters associated with the implementation of RCCI as an advanced clean and efficient combustion strategy.

1.1.5 Potential of RCCI strategy to achieve targets of BSVI emission norms, their limitations, andpossible solutions

2. Historical Review of RCCI

As an effect of simultaneous reductions in the NOx as well as PM emission of the diesel engine without affecting the engine efficiency and the original performance, various low temperature combustion (LTC) strategies have been developed. Amongst various low-temperature combustion strategies, HCCI (homogeneous charge compression ignition), PCCI (premixed charge compression ignition), RCCI are important low-temperature combustion strategies.[9,10,18,37,45,46] The common features of all the LTC strategy are higher degree of charge homonization and overall leaner operation.[1,14] Before RCCI, worldwide researchers have been studied HCCI and PCCI for decades.[32,55] In the case of HCCI combustion control canbe carried out by managing the charge condition at the closing of the inlet valve. This was highlydependent on the operating conditions; this was not useful for the full load conditions of the engine.[1,45] Since HCCI gives higher thermal efficiency and reduction in emissions, it has certain issues like shorter combustion duration which is responsible for the high peak pressurerise and develops the problem of noise, vibration and harshness especially due to higher peaks of the pressure rise rates as an effect of advanced injection effects.[45,55]To overcome this issueToyota et al proposed premised diesel combustion mode (diesel PCCI). Higher rate of EGR utilized in this strategy to prolong the ignition delay to enhance the mixture hominization.[28] Ahigher degree of homonization and overall leaner effect reduces the PM and NOx simultaneously.[32] But the combustion stability affected at higher load as an effect of a higherrate of EGR. For limited load capacity, the early injection strategy improves the combustion stability but this will affect the thermal efficiency.[45] In the case of the PCCI, due to stratification of the charge in the combustion chamber and staged combustion with reduced pressure rise rate, it gives wider range than HCCI but the issue was controlling the combustion and mechanical stresses as an effect of the auto ignition qualities of the diesel fuel. To overcome these problems; kalghatgi et al [28] introduced the methodology of gasoline PPC (partial premixed charge compression ignition) in 2006. Use of fuel with auto ignition temperature higher than diesel allows a large ignition delay period and there is no need for too much EGR fractions, too early injection before TDC, too low CR (compression ratio), or dedicated hardware.As an effect of availability of larger mixing time it leads to lower NOx and PM as in comparison with the conventional diesel combustion process. For lower load operation, gasoline with octane number greater than 90, there were some difficulties, these difficulties tried to overcomeby double injection and spark assistance but it leads to a large amount of NOx and PM generation.[41] So the experimentations carried out by the Bessonette et al.[6] concluded that the fuel characteristics can be varied according to the load conditions of the engine for low load operations diesel like fuels can be used while for high load operations gasoline like fuels can beused.

After that Inagaki et al.[27] introduced the concept of the dual fuel combustion, in which there is a port fuel injection of the is octane while the direct injection of the diesel fuel. They reported that there is better control over the combustion process with very low NOx and PM emission. Later on,bessonette et al[6] found that the proper fuel for the HCCI operation having reactivity lies in between the diesel and gasoline. Based on these research and findings, kokjohnet al [47] had conducted various experiments for the duel fuel PCCI and HCCI by varying the reactivity, with a combination of the EGR for the control of combustion with maximum efficiency. They reported that an adjustment in the reactivity of the fuel leads to optimum efficiency in certain operating conditions. In these experimentations there were port injection ofthe gasoline as a low reactivity fuel (LRF) and direct injection of diesel as a high reactivity fuel (HRF) in the combustion chamber. This phenomenon is called as reactivity controlled compression ignition.[18,45,47,52,12]

3. Fundamentals of RCCI Strategy

The control of fuel reactivity inside the combustion chamber is an important feature of the RCCI.[18, 45, 52,61]By its design, RCCI is a combustion strategy that uses two fuels having different reactivity utilized for fuel mixing in the cylinder. The principle of RCCI operation is as shown infigure 1.[47]

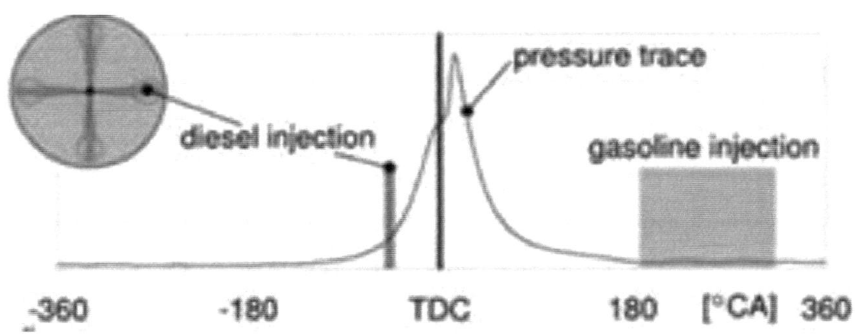

Figure 1 Fundamentals of RCCI operation

Low reactivity fuel is of gasoline category (i.e. low cetane number) is injected too early by port fuel injection (PFI) and high reactivity fuel in the category of diesel fuel (i.e. high cetane number) is directly injected single or multiple times during the combustion cycle.[30]Subsequently, the reactivity stratification is established. The fig.1 shows the working principle of the RCCI engines.

3.1. RCCI Chemistry

In RCCI combustion strategy 80 to 90 % of low reactivity fuel like gasoline can be transferred through the port while the direct injection of the remaining amount of the high reactivity of thefuel like diesel in the combustion chamber. As an effect of the reactivity gradient it has the potential to reduce the NOx as well as PM even without too much necessity of the EGR(Exhaust gas recirculation) at higher load operations.[45] The combustion process in the RCCI engines mainly depends on the mixture quality of the charge. The combustion process in the RCCI engineis a staged combustion process which extends premixed combustion duration, resulted in higherthermal efficiency with lower emissions as well as, low rate of pressure rise at higher load operation of the engine.[39]

There are three phases of the combustion in RCCI engines are i) Low temperature combustion flame i.e. ignition of the low reactivity regions ii) ignition of low reactivity zones lead to a higher heat release rate as an effect of high mass of premixed charge due to multiple site combustion iii) Generation of the multiple flames due to multiple stage combustion.[4] In rcci, the combustion process initiates from the 10 to 20% of the HRF supplied by the single or double injections and HRR can be controlled by the spray formation process.[52]

3.2. Heat release rate

RCCI consists two stage heat release process. The first stage is known as LTHR (low-temperatureheat release) while second is HTHR (high-temperature heat release).

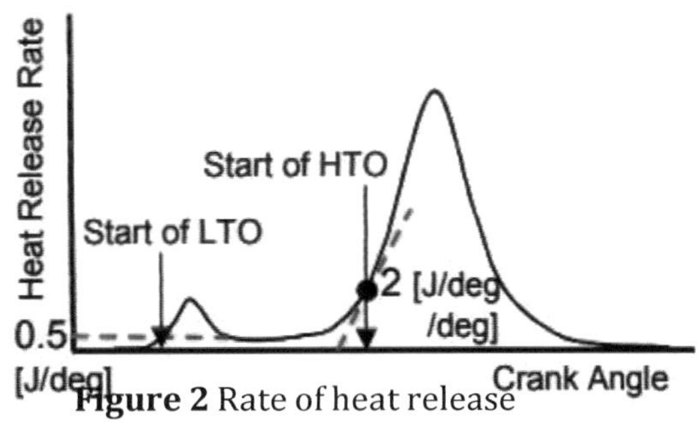

Figure 2 Rate of heat release

Figure 2 shows the rate of heat release of the RCCI operation in terms of start of high temperature oxidation(HTO) and low temperature oxidation(LTO).[33] The maximum rise of LTHR is mainly controlled by the high reactivity fuel with NTC (negative temperature coefficient). The HTHR is mainly dependent on the quantity of the premixed charge, i.e., LRF (low reactivity fuel) in RCCI combustion.[33]

4. BSVI Emission Regulations

To get rid from consequence of these exhaust emissions, various emission regulations were implemented throughout the world. India have their emission regulations known as Bharat stage emission standards, likewise euro norms for European countries, Canadian Environmentalprotection act for Canada, Environmental protection agency, United States of America, China Energy policy of China, China has their emission norms.[45,54] Emission regulation or the legislation had set the limits of the exhaust constituents. Table 1 shows the Bharat Stage VI (BSVI) emission norms for the passenger as well as heavy commercial vehicles.

In India, from April 1, 2020, implementation of the Bharat stage 6 (BSVI) norms throughoutthe country.[23] The target of emission norms BS-VI is to maintain the PM at 0.01 g/kWh and NOx at 0.4 g/kWh levels. India already skipped the Bharat Stage emission norms (BSV) norms, so there is a drastic change in the limit of the exhaust constituents from BSIV to BSVI. For BS-VI, it is proposed to reduce NOx by 89% and PM by more than 50% about the BS-IV limits in case of heavy commercial vehicles.

Table 1 BSVI Norms [23]

Emission standards for passenger vehicles-Diesel						
Pollutant Types(In g/Km)						
Stage	CO	HC	HC+NOx	NOx	PM	PN*^
BS-IV	0.50	-	0.30	0.25	0.0250	-
BS-VI	0.50	-	0.17	0.08	0.0045	6.0x10^11
Emission standards for heavy commercial vehicles (GVW>3.5T)						
Pollutant Types(In g/Km)						
Stage	CO	HC	HC+NOx	NOx	PM	PN*^
BSIV(ESC)	1.50	0.46	-	3.50	0.02	-
BSIV(ETC)	4.00	055	-	3.50	0.03	-
BSVI(WHSC)	1.50	0.13	-	0.40	0.01	8.0x10^11
BSVI(WTSC)	4.00	0.16	-	0.46	0.01	6.0x10^11

ESC- European Steady Cycle; ETC – European Transient cycle WHSC- World Harmonised Stationery Cycle; WSTC- World Harmonised Transient CyclePN count for WTSC will be applicable from April 2025.

4.1Technological transformation review for BSVI and its economical impacts

Emissions regulations in the Bharat Stage VI (BS-VI) pointed out for an after-treatment exhaust system (EATS) with very high PM &NOx mitigation capabilities and the mitigation in PN. All thetechnologies are associated with the in cylinder technologies and after treatment systems.[45]

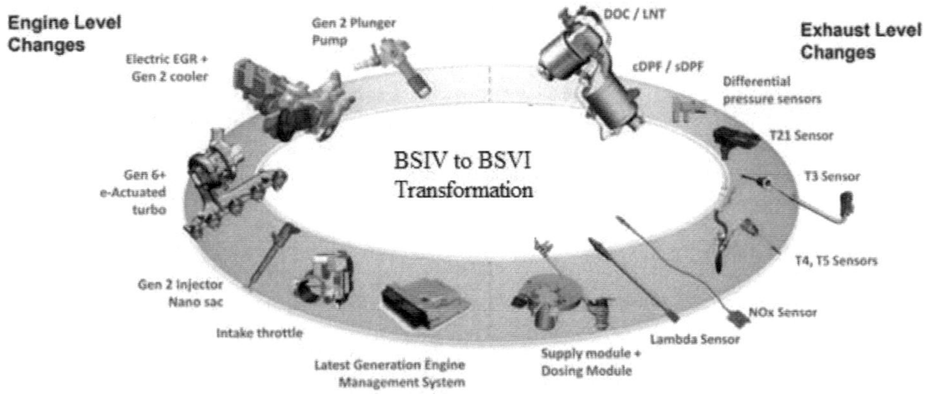

Figure 3 Changes in Engine technology BSIV to BSVI

Fuel modification, engine design modifications, exhaust after-treatment, optimization of the existing systems are the various ways to regulate the emission from the engine. As the economics of the transformational changes in the BSIV to BS-VI is a concern, there are an increase in the price from 10 to 15% of the existing cost of the diesel operated vehicle as per the SIAM (Society of Indian automotive manufacturers) reports [24,22] table2 shows segment-wise cost implications of the BSVI technology.[25]

To shift the existing engine technology from BSIV to BSVI it is essential to change the existingtechnologies of engine level as well as exhaust level as shown in figure 3. As well as table 3 technological as well as fuel level changes required for BSVI Transformation.

Table 2 Cost Implications of BSVI Technologies [25]

Segment	**Technology required to meet BS-VI Norms**	**Increase in cost**
Diesel operated passenger vehicle	Main changes in fuel injectionsystem LN T DP F	Vehicle cost my increase by 10 to 12%
Light & MediumCVs	High Pressure Direct InjectionDOC ,EGR + SCR ,DPF	Increase in engine cost ~50% from abase level of BSIV category of same capacity
Heavy CVs	high Pressure Direct InjectionDOC EGR + SCR ,DPF	Increase in engine cost ~50% from a base level of BSIV category of samecapacity

Table 3 Technology up gradation BSIV to BSVI [55]

Technologies	From BS IV to BS VI transformation	
Fuel utilized	BS IV fuels	BS VI fuels

Combustion	Advancements in combustion an d calibration process tocontrol the PM Use of VGT with intercooling -NOx emissio ncontrollers	Advancements in combustion and calibration ofthe engine - advancements in fuel injection systems - advanced VGT Advanced No controlled and ElectronicsPrecise control on EGR rate.
On-board diagnostics	OBD Stage I	Addition of following in OBD Stage II More advanced ECU and engine managementsystem Use of standardized OBD systems and their remoteaccess for maintenance Requirements for additional monitoring systemsfuel injection Exhaust Gas RecirculationBoost EGR cooling
After-treatment system	Open loop SCR for NOxreduction -PM control (for EGR pathway only)	Closed loop SCR system for the NOx reduction,LNTs -PM control: DOC + DPFs(cDPF and sDPF)

5. Implementation of RCCI Combustion

5.1. Fuel Management of RCCI engine

Fuel stratification plays a key role in the case of RCCI combustion characteristics and overall performance of the engine.

5.1.1 Single Fuel Strategy with additives

In a single fuel RCCI strategy only LRF will be used, cetane improver additives are used to develop the reactivity gradient in the combustion chamber and LRF directly injected in the combustion chamber. [52] In this strategy a single fuel tank is sufficient for the operation. In thissingle fuel strategy, Gasoline, ethanol, methanol, iso-butanol can be used as an LRF for the RCCI combustion. [52] The fuel with cetane improver additives works as an HRF. 2-EHN (Ethylhexyl nitrate) and DTBP (Di tertiary butyl peroxide) have been used for the RCCI combustion.[33,59]

5.1.2 Double Fuel Strategy

Initially, various studies were conducted by using gasoline as HRF and Diesel as LRF. With theseconventional fuels different types of renewable fuels like biodiesel, alcohol, natural gas, etc wereused for the RCCI combustion. Alcohols and NG creates larger reactivity gradient as compared togasoline and they can extend the operating loads also. Alcoholic fuels reduce the NOx formationbecause of its oxygen enrichments.

5.2 Engine Management of RCCI Engines

While adopting the RCCI combustion technology various parameters need to be optimized like injection strategy, SOI (Start of injection) timing, injection duration, spray angle, fuel injection pressure and fuel ratio, piston bowl geometry, etc. These parameters directly affect the combustion characteristics and overall performance of the engine.

5.2.1 Fuel Ratio

Fuel ratio can be defined as the ratio of LRF to HRF in mass, energy or volume, etc. In cylinder, reactivity can be controlled with the help of fuel ratio. Hanson et al[20] conducted experimentation on a heavy-duty single-cylinder engine to study the effect of varying the fuel ratio of gasoline percentage of 82, 86,89% at constant IVC timing 143^0 ATDC on performance and emission characteristics. They reported that a higher percentage of gasoline retards the combustion phasing as an effect of global fuel reactivity.

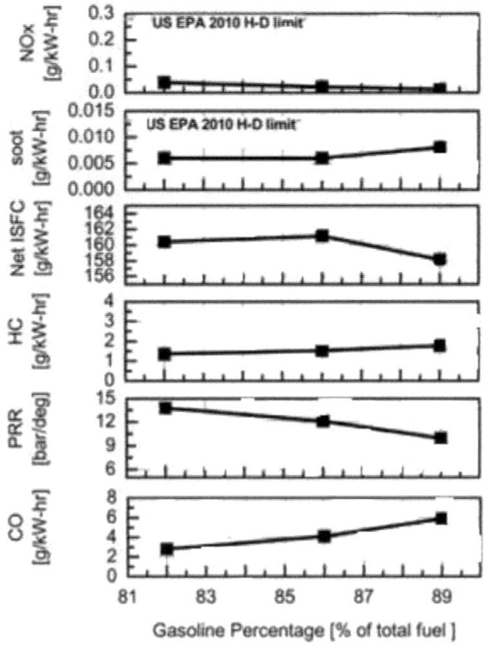

Figure 4 Effect of gasoline percentage on emission characteristics

This retarded combustion phasing leads to lower heat transfer as well as drops the combustion chamber temperature and allowed for the lowest fuel consumption as well as enhances the indicated thermal efficiency. It also reduces NOx and PM emissions. Figure 4 showsthe effect of gasoline percentage on the emission characteristics of the engine.[20]

5.2.2 Effect of first and second injection timings

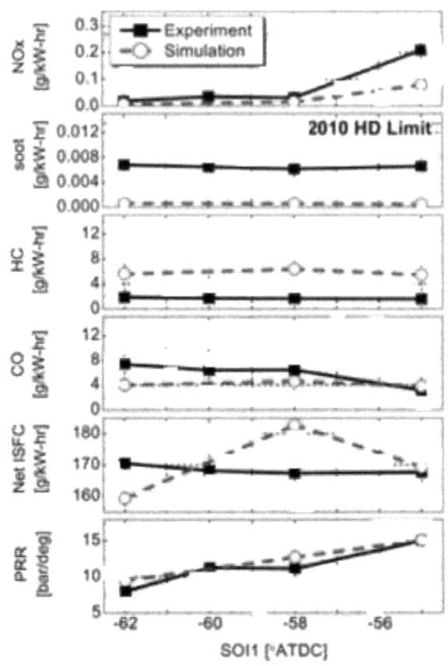

Figure 5 Effect of SOI1 timing on performance and emission characteristics

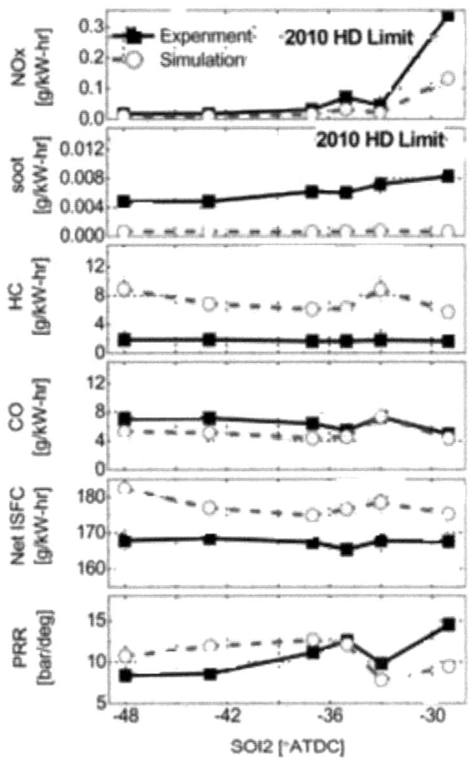

Figure 6 Effect of SOI2 timing on emission characteristics

There are various controllable parameters in the RCCI combustion process, the start of combustion is one of them because it is closely associated with the mixing process and combustion kinematics. Both the early, middle, and late injection timing effects on combustion characteristics and reflects in the HRR and the thermal efficiency of the engine. Hanson et al.[20] tested heavy-duty engine for 9 bar net imep and 1300 rpm for the injection timing. The fuel supplied through the port fuel injection systems 73% while 27 % diesel through the direct injection system about this 62% quantity of the diesel in the first injection while remaining 28
% fuel in the second injection.Figure 5 shows that as advancement in the engine timing (SOI1) retards combustion phasing resultedin peak pressure rise. [20] The NOx reduced and but CO increased and not too much change was observed in the HC emission. Figure 6 shows experimental and simulated results of and rate of heat pressure inside the cylinder and heat releaserate over a SOI2 timing.[20] Retardation ofsecond injection leads provides very less timefor the preparation of the charge, this improper mixing generates the local cylinder temperature, increases the NOx and PM. There was no effect on the fuel consumption as well as CO. Advanced SOI2 decreases the pressure rate as a result of homogeneous mixture and delayed combustion.

5.2.3 *Mass fraction in the first and second injection*

The mass fraction of the split injection depends on the SOI timing of each injection pulse. Hansonet al[20] conducted an experiments to investigate the influence of mass fraction in the first andsecond injections of HRF.

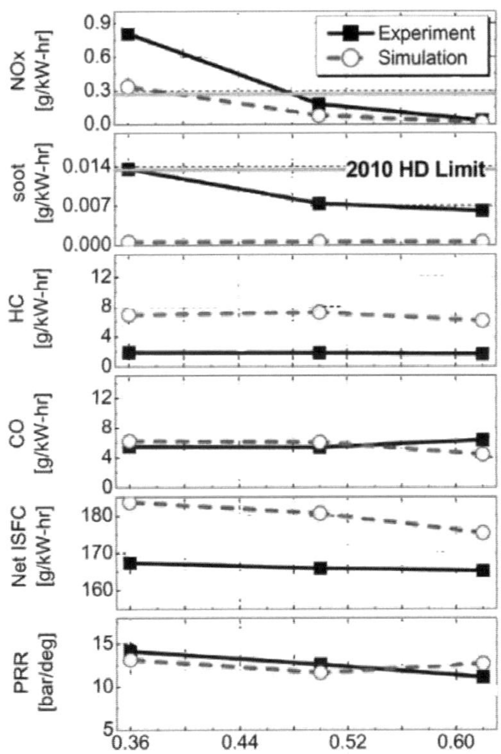

Figure 7 effect of mass fraction on emission characteristics

The mass fraction variation of the diesel fuel injected in SOI1 was from 36% to 62%of the totaldiesel fuel injected and it was 27% of the total fuel supplied to the combustion chamber in each stroke. For the addition of second injection fuel, NOx and PM have major rises. In the case of later injection of the diesel fuel in the process, results in more stratification of the charge, and increases the reactivity of the charge in the cylinder. These rich and high reactive zones are starting to ignite higher cylinder temperatures earlier, causing a rise in NOx and PM. Figure 7 shows effect of the fraction of diesel in each pulse on performance and emission characteristics of the engine and corresponding operating conditions of the engine.[20]

5.2.4 Spray Angle

Spray angle performs a key role in the overall combustion process of the RCCI engine. Properspray angle effects on the distribution of the fuel particles in the combustion chamber and this distribution directly effect on the combustion, performance and emission characteristics of the engine. Spray angle can be defined as the angle between the cone axis of the diesel fuel spray andthe axis perpendicular to the piston crown. The spray angle is associated with how much fuel isreaches the location of the squish region of the combustion chamber. Nazemi et al. [42] investigated the effect of various injection parameters on the emission, combustion and performance characteristics of the RCCI engines, they selected. They concluded that spray angleis the important parameters of the injection design which affects on the combustion phasing andalso in the heat release rate.

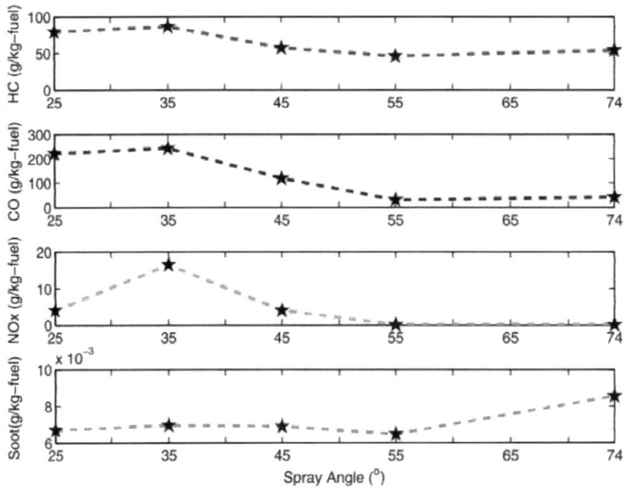

Figure 8 Effects of spray angle on emissions characteristics

In the RCCI strategy the diesel fuel injected far early before the top dead centre so part of thefuel injected is impinged the wall of the cylinder and it makes the fuel wall on the cylinder wall.Part of the fuel undergoes in the squish region. Figure 8 shows the effect of spray angle on emission characteristics of the engine.[42]

5.2.5 Injection pressure

Homonization, vaporization and dispersion of diesel fuel are the important sub processes of thein cylinder mixing process and that can be affected by the injection pressure.

There are two aspects of the injection pressure one is, if the injection pressure is high,

as a result of a higher rate of vaporization the mixture becomes homogenous and another hand if the particle size reduces the inertia of the particles reduces and reduces the penetration lengthwise. For the same amount of the injected fuel, as the injection pressure increases there is a reductionof injection duration. Higher injection pressure responsible for rapid combustion rates, which resulted in the higher in cylinder temperature. The reduction of the torque as well as power is due to the inefficient transfer of the heat into useful work. The fundamental reason is that the combustion starts in a very small area where the fuel is injected and the dispersion of the flames around the cylinder. During this travel there was not proper utilization of the air.
Nazemi et al [42] have conducted experiments to explore the results of varying injection pressures from 380 bar to 580 bar in the 20 bar range for the 4-cylinder diesel engine. Increasedinjection pressure from 380 bar to 580 bar was reported to cause a decrease in peak pressure byapproximately 10 bar. Figure 9 shows effects of injection pressure on emissions including HC, CO, NOx and soot [42]

The basic cause of this, either raising the injection pressure or reducing the duration of the injection, the fuel parcel diameter decreases and these parcels of smaller diameter unable to distribute as per requirements and deposited together in the high-temperature piston bowl areaIneffective self-ignition would also occur, which cannot spread through the entire combustion chamber and which will provide sufficient energy rapid burning of the fuel. The peak HRR decreases by increasing the injection pressure and combustion is progressively delayed. An increase in injection pressure reduces the rate of auto ignition process near the tank of the piston, as less gasoline fuel burns compared to low injection pressure and thus decrease in the peak of the HRR. As a result of higher injection pressure and lower in cylinder gas temperature there were a reduction in the NOx and quite higher values of HC, CO, and soot had reported.

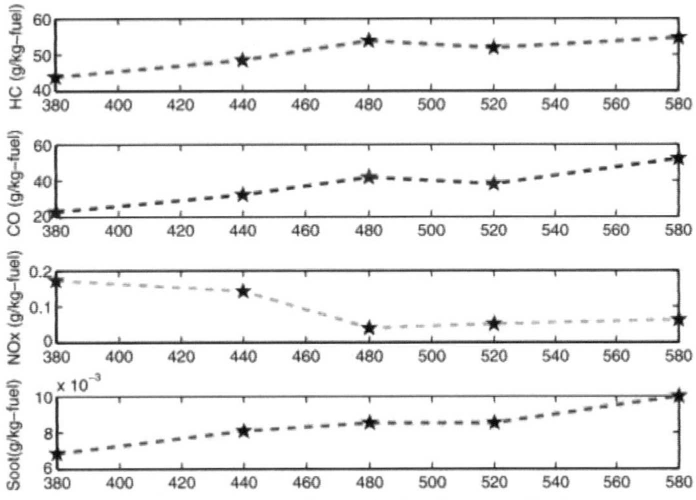

Figure 9 Effects of injection pressure on emissions including HC, CO, NOx and soot

5.2.6 EGR Rate

Exhaust gas recirculation is one that contributes to the reduction of NOx emission and PRR. Durasami et al [16] conducted experimentation to explore the effect of the EGR on combustion,performance as well as emission characteristics of the methanol/diesel and methanol/PODE RCCI Combustion process for the automotive engine.

They reported that use EGR resulted in a reduction of the peak in-cylinder pressure as well asHRR for both Methanol/Diesel and Methanol/PODE RCCI combustion due to the effect of charge dilution and heat capacity of the EGR. EGR delayed the start of combustion as well as suppressedthe cylinder pressure up a certain limit, if the limit exceeds the engine may lead to misfiring.

Ultimately they reported that the combustion phasing and extension of the load range associatedwith the amount of EGR. As emission is concerned, reduction in the oxygen concentration as aneffect of the EGR leads to an increase in the soot, BSHC, and BSCO emissions but the notable reduction in NOx for RCCI combustion with EGR in comparison with without EGR models shownin figure 10. [16]Rate of pressure rise reduced in both cases for EGR in comparison with withoutEGR which is helpful to enhance the operating range.

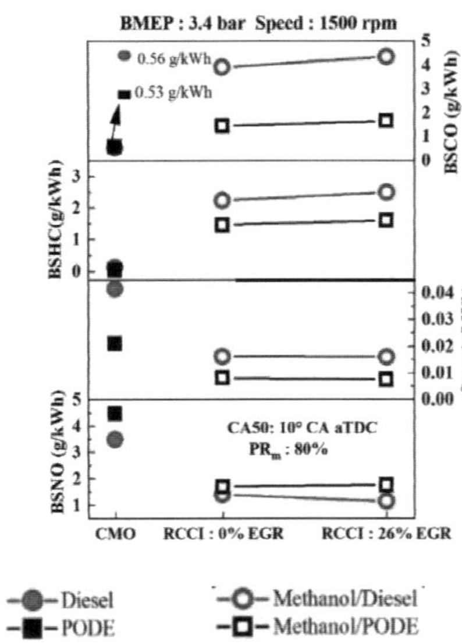

Figure 10 Effect of EGR on performance on emission characteristics of the engine

Table 4 Summary of Various Important Parameters of RCCI Emission Characteristics

Sr.No.	Parameter	Overall conclusion	References
1	Pre mixed ratio	As the gasoline fuel percentage increases there were increases in HC, CO, soot, and reduction NOx. This is because of the reduction in the incylinder gas temperature with increasing gasoline fraction.	Hanson et al. [20]Kokjohn et al.[31]Ma et al.[40]Benajes et al. [4]Dempsey et al. [10]Cha et al.[7]Li et al. [38]Li et al. [34]

2	First an dsecond main injection timing	The heat release is a single-stage process for the early start of injection and a two-stage heat release process for late start of injection , increase in pressure rise rate has been observed increases in temperature of the cylinder which increases the NOx but reduces the HC and CO emission.	Ma et al. [40]Benajes et al.[4]Desantes etal. [13]Li et al. [38]
3	The mass fraction in First and second injection	Effect of split injection mass fraction fundamentally depends a lot on the SOI timing of each injectionpulse. Split injection fractions may affect the location where the fuel has injected either bowl or squish of the combustion process in the cylinder, and therefore there is an increase in PPRR and ringing intensity. But there was no noticeable impact on the performance and emission characteristics of the engine.	Ma et al. [40]Paykani et al[48] Nieman et al[43]
4	Spray angle	In case of a broader spray angle, fuel burning happens in the area of centreline and squish respectively, and in case of narrower spray angles, itallows the charge to burn close to the cylinder bulk, which produces higher levels of HC and CO emissionsnear the area of cylinder wall as well as crevice volumes. Also, the emission of NOx is higher with a lower spray angle as a result of higher temperaturesand rich and locally high reactive zones.	Amani et al [3],Poorghase mi et al [49] Khatamnejad et al[29]
5	Injection Pressure	For the same amount of the fuel, if injection pressureis higher than the duration of injection reduces. Up to a certain limit of injection pressure provides betteratomization results in reduction HC and CO emission	Ramasankaran et al[50] Duraisamy et al[18]
6	EGR Rate	An effect of dilution of the charge and thermodynamic effects of EGR reduces the temperature of the combustion chamber. There is a reduction in the NOx formation. In some cases, increase in CO and HC was observed as an effect of reduction in oxygen availability in the combustion chamber.	Jing Li et al (36)Ma et al.[40]Wu et al[62]

6. Applications of RCCI to Meet BSVI Emission Norms

Combustion optimization of the diesel engine is essential for regulating the emission constituents from the exhaust gases, since the RCCI combustion having good control over the combustion event which leads to the drastic reduction in the NOx and PM. Various factors directly affect the NOx formation process, which includes type of fuel used, fuel injection pressure, injection timing, pulses of injection, amount of fuel injected in each pulse, premised fuel ratio, piston bowl geometry, etc.[1]

Benajes et al [3] experimentally investigated the capability of the RCCI engine to meet the EURO VI emission norm. They tested single cylinder medium duty engine developed for the urban freight transportation. The engine was modified as per the requirement of the RCCI operation. They used three combustion strategies based on the loading condition. Upto 40% there was fully premixed strategy, for 40 to 75% highly premixed, and for 75 to full loading condition the strategy was diffusive. To achieve the EURO VI They followed three steps stable engine operation, the optimistic performance of the engine for NOx parameters, and third for optimization for the HC and CO emission. As NOx emission is concerned it was observed that theshape of the NOx emission map is closely associated with the gasoline mass fraction and the rateof EGR. As the fraction of the diesel fuel increases the locally rich mixture zones and higher temperature which leads to the generate more NOx.as a final result was concerned there is reduction of NOx ranging from 0.6 and 0.8g/kWh and the some peaks of the about the 1.2 to 1.4g/kWh which is typically less than the current diesel engines which operates in full load conditions with SCR. The overall reduction in the NOx varies in the range of 75 to 98% in comparison with CDC operation. So there was no need to use the reductant like urea in SCR to achieve the targets set by the BSVI emission norms,because the engine operation shows the maximum rise of the NOx reduction in the region of low speed, as shown in figure 11.[3]

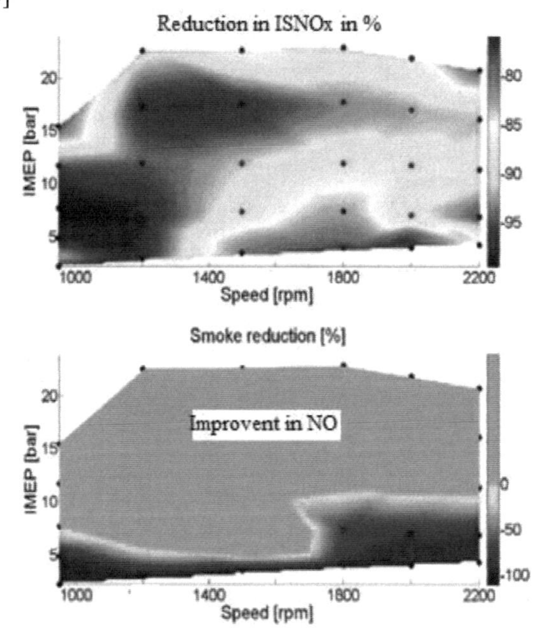

Figure 11NOx and smoke emissions reduction with the dual-fuel strategy versus CDC

For the HC and CO emissions are concerned as load and speed of the engine increases both the unburned hydrocarbon HC and CO increases. For idle and low speed engine operation i.e forthe speed range 950-1500 rpm the maximum rise for both the HC and CO pollutants was reported. The combustion strategy was followed in such a way that the temperature of the gas ranges from 1500-1700K, where complete oxidation of the CO lower limit and there were formation of the minimum amount of the NOx . for above 10 bar IMEP, a very small amount of HC emissions were observed, a for the mixture strength is concerned equivalence ratio rangesfrom 0.7 and 0.8 and gas temperatures about 1700 K without maximum fuel consumption, theCO and HC emission is less with

NOx and ultra-low PM contents, pressure and temperature zones as a source of ignition. It can be concluded that Both the CO and HC emissions are the functions of both equivalence ratio as well as gas temperature. There was very little time available for the mixing process and maybe the chance of the soot formation but the injection quantity of the diesel fuel, the exhaust emissions are meets the targets of Euro VI/ BSVI normsfor all speed and 75% load range. Similarly Xu et al [63] investigated the use of variable valve timing (VVT) in RCCI combustion for combustion, performance, and emission characteristics ofthe heavy-duty engine for wide range of the load. They optimized various parameters like premixed ratio, intake valve closing (IVC) timing, injection timing, and quantity of EGR, intake airpressure, and temperature. The investigations aimed to meet the Euro 6 emission regulation without using exhaust after treatment systems without affecting the thermal efficiency of the engine. They concluded that by optimization of RCCI combustion with the help of VVT strategy, the Euro6/BS6 NOx targets were achieved with ultra-low soot emissions at low and medium load range, but for the higher load operation requirement of at least one after treatment devicefor further elimination of the NOx or soot emissions.

Similarly Xu et al [63] investigated the use of variable valve timing (VVT) in RCCI combustionfor combustion, performance, and emission characteristics of the heavy-duty engine for wide range of the load. They optimized various parameters like premixed ratio, intake valve closing (IVC) timing, injection timing, and quantity of EGR, intake air pressure, and temperature. The investigations aimed to meet the Euro 6 emission regulation without using exhaust after treatment systems without affecting the thermal efficiency of the engine. They concluded that byoptimization of RCCI combustion with the help of VVT strategy, the Euro6/BS6 NOx targets wereachieved with ultra-low soot emissions at low and medium load range, but for the higher load operation requirement of at least one after treatment device for further elimination of the NOxor soot emissions.

7. Research directions and future scope

- The key challenge of the RCCI combustion is to control the timing and magnitude of HRR for higher loading. Further studies are required to explore the various options like DDFS and MPCI, multiple pulses, etc
- In the case of RCCI combustion, normally too early injection is used to develop required reactivity stratification. But early injection of the fuel may lead to fuel droplets adherence onthe cylinder wall or called asfuel impingement. This results in incomplete combustion of thecharge.To mitigate this more studies are required to minimize the effects of fuel deposition inthe crevice areas with the help of blends of oxygenated fuels, spray angle, spray patterns, noof holes ,injection pressure, and their cumulative effects on the combustion , performance, and emission characteristics of the engine
- It is essential to minimize the effects of the HC and CO through a specially designed after- treatment system with low temperature catalyst or there should be provision to increase temperature to perform required catalytic reactions or there may need for fuel born catalyst.
- There is need to develop effective control models on RCCI combustion as far as commercial use of these strategies is concerned since the vehicular traction requirements continuously changes. As compared to the CDC, RCCI engine is limited by both at high and low operating range. At the high loads, there is an issue of high levels of PPRR and combustion noise. At lowload CDC gives similar efficiencies and

NOx emission with lower UHC and CO emissions in comparison with RCCI engines. Mode switching as per the required load will be the solution for this and efforts are required in this direction.

8. Conclusions:

To achieve the emission regulations prescribed in the BSVI norms both the engine and exhaust level changes are carried out by the various engine manufacturers. Advanced engine combustionprocess RCCI has a great potential to get higher thermal efficiency as well as reduce the NOx andPM emissions as prescribed limits of the BSVI emission norms as an effect of the higher degree of homonization and effective control through the reactivity stratifications. RCCI will provide thecost-effective solution for emission reduction which reduces the economic burden on the enginemanufacturer as well as the owner.

- Advancements in the diesel injection timing, there was better stratification of the mixture andthere are few areas of lower local reactivity across the cylinder. That means reducing the firstpeak of HTHR and raising the second peak.
- RCCI combustion has the potential to reduce the soot and NOx, in comparison with CDC which operates on neat diesel, which shows the ability to achieve the targets, set by the BSVIemission norms. On the downside, higher level CO and unburned HC were assessed, the mainreason behind this is fuel deposited in the crevice areas, and where the burning of the fuel isquite difficult.
- Optimized engine and fuel management system of the RCCI engine strategy can be used to achieve wider load range of the operation with higher thermal efficiency and avoids costly after treatment system or there will be minimum requirements of the exhaust after treatment. This will directly reduce the initial as well as the running cost of the vehicle.
- The research reveals that current research is limited with the hybrids vehicles and is in the very initial stage, further modifications like advancement in the engine control as well as engine management systems; changes required to fulfill the traction requirements of the vehicle in various operating conditions like mode swing technologies to provide the requiredmixture requirements.

References

[1] Agarwal A K, Singh A P and Maurya R K 2017 Evolution, challenges and path forward for low temperature combustion engines *Prog. Energy Combust. Sci.*61 1–56

[2] Amani S M E and Chitsaz R H I 2019 Effects of the swirl ratio and spray angle on the mixture stratification in a diesel – NG RCCI engine *J. Brazilian Soc. Mech. Sci. Eng.*5

[3] Benajes J, García A, Monsalve-Serrano J and Boronat V 2017 Achieving clean and efficient engine operation up to full load by combining optimized RCCI and dual-fuel diesel-gasoline combustion strategies *Energy Convers. Manag.*136 142–51

[4] Benajes J, Molina S, García A, Belarte E and Vanvolsem M 2014 An investigation on RCCI combustion in a heavy duty diesel engine using in-cylinder blending of diesel and gasoline fuels *Appl. Therm. Eng.*63 66–76

[5] Benajes J, Pastor J V., García A and Monsalve-Serrano J 2015 The potential of

RCCI concept to meet EURO VI NOx limitation and ultra-low soot emissions in a heavy-duty engine over the whole engine map *Fuel*159 952–61

[6] Bessonette P W, Schleyer C H, Duffy K P, Hardy W L and Liechty M P 2007 Effects of fuel property changes on heavy-duty HCCI combustion *SAE Tech. Pap.*2007 776–90

[7] Cha J, Kwon S, Kwon S and Park S 2012 Combustion and emission characteristics of a gasoline –dimethyl ether dual-fuel engine 226 1667–77

[8] Dec J E 2009 Advanced compression-ignition engines - Understanding the in-cylinder processes *Proc. Combust. Inst.*32 II 2727–42

[9] Delvescovo D 2018 The Effects of Charge Preparation , Fuel Stratification , and Premixed Fuel Chemistry on Reactivity Controlled Compression Ignition (RCCI) Combustion

[10] Dempsey A B, Curran S and Reitz R D 2015 Characterization of Reactivity Controlled Compression Ignition (RCCI) Using Premixed Gasoline and Direct-Injected Gasoline with a Cetane Improver on a Multi-Cylinder Engine *SAE Int. J. Engines*8 859–77

[11] Dempsey A B, Walker N R, Gingrich E and Reitz R D 2014 Comparison of low temperature combustion strategies for advanced compression ignition engines with a focus on controllability *Combust. Sci. Technol.*186 210–41

[12] Dempsey A B, Walker N R and Reitz R 2013 Effect of Cetane Improvers on Gasoline, Ethanol, and Methanol Reactivity and the Implications for RCCI Combustion *SAE Int. J. Fuels Lubr.*6 170–87

[13] Desantes J M, Benajes J, García A and Monsalve-Serrano J 2014 The role of the in-cylinder gas temperature and oxygen concentration over low load reactivity controlled compression ignition combustion efficiency *Energy*78 854–68

[14] Dev S, Chaudhari H, Gothekar S, Juttu S, Harishchandra Walke N and Marathe N V. 2017 Review on Advanced Low Temperature Combustion Approach for BS VI *SAE Tech. Pap.*2017-Janua

[15] Dong Y 2018 Reactivity Controlled Compression Ignition (RCCI) Combustion Using Methanol and Diesel in a Single Cylinder Research Engine

[16] Duraisamy G, Rangasamy M and Govindan N 2020 A comparative study on methanol/diesel and methanol/PODE dual fuel RCCI combustion in an automotive diesel engine *Renew. Energy*145 542–56

[17] Feng H, Zhang C, Wang M, Liu D, Yang X and Lee C F 2014 Availability analysis of n-heptane/iso- octane blends during low-temperature engine combustion using a single-zone combustion model *Energy Convers. Manag.*84 613–22

[18] Ganesh D, Ayyappan P R and Murugan R 2019 Experimental investigation of iso-butanol/diesel reactivity controlled compression ignition combustion in a non-road diesel engine *Appl. Energy*242 1307–19

[19] Gao T 2017 Scholarship at UWindsor MIXING AND REACTIVITY CONTROL WITH BUTANOL AND ETHANOL IN COMPRESSION IGNITION ENGINES

[20] Hanson R M, Kokjohn S L, Splitter D A and Reitz R D 2010 An Experimental Investigation of Fuel Reactivity Controlled PCCI Combustion in a Heavy-Duty Engine 3 700–16

[21] Hodges K A, Aniello A, Krishnan S R and Srinivasan K K 2017 Impact of propane energy fraction on diesel-ignited propane dual fuel low temperature

combustion *Fuel*209 769–75

[22] https://economictimes.com/industry/auto/cars-uvs/as-indiswitches-to-bs-vi-sub-rs-10-lakh- petrol-cars-may%20gain/articleshow/73266356.cms?from=mdr

[23] https://morth.nic.in/sites/defult/files/AIS_137_%28Part_4%29_regarding_BS_VIemission _Norms.Pdf

[24] https://timesofindia.indiatimes.com/business/india-business/bsvi-shift-may-hike-princees-up- to10-siam-auto-news/articleshow/73203904.cms

[25] https://www.icra.in/Files/Articles/SH-2016-Q2-6-ICRA Passenger%20vehicles.pdf

[26] https://www.mahindra.com/resources/investorreport/FY20/Events%20 Update/BS-VI-Knowledge-Sharing-Session-Presentation-3rd-June 2019.Pdf

[27] Inagaki K, Fuyuto T, Nishikawa K, Nakakita K and Sakata I 2006 Dual-fuel PCI combustion controlled by in-cylinder stratification of ignitability *SAE Tech. Pap.*2006

[28] Kalghatgi G T, Risberg P and Ångström H E 2007 Partially pre-mixed auto-ignition of gasoline to attain low smoke and low NOx at high load in a compression ignition engine and comparison with a diesel fuel *SAE Tech. Pap.*

[29] Khatamnejad H, Khalilarya S and Jafarmadar S 2018 Influence of Blend Ratio and Injection Parameters on Combustion and Emissions Characteristics of Natural Gas-Diesel RCCI Engine

[30] Khodadadi Sadabadi K, Shahbakhti M, Bharath A N and Reitz R D 2016 Modeling of combustion phasing of a reactivity-controlled compression ignition engine for control applications *Int. J. Engine Res.*17 421–35

[31] Kokjohn S L, Hanson R M, Splitter D A and Reitz R D 2010 Experiments and modeling of dual-fuel HCCI and PCCI combustion using in-cylinder fuel blending *SAE Int. J. Engines*2 24–39

[32] Krishnamoorthi M, Malayalamurthi R, He Z and Kandasamy S 2019 A review on low temperature combustion engines: Performance, combustion and emission characteristics *Renew. Sustain. Energy Rev.*116 109404

[33] Li J, Yang W and Zhou D 2017 Review on the management of RCCI engines *Renew. Sustain. Energy Rev.*69 65–79

[34] Li J, Yang W M, An H and Zhao D 2015 Effects of fuel ratio and injection timing on gasoline/biodiesel fueled RCCI engine: A modeling study *Appl. Energy*155 59–67

[35] Li J, Yang W M, An H, Zhou D Z, Yu W B, Wang J X and Li L 2015 Numerical investigation on the effect of reactivity gradient in an RCCI engine fueled with gasoline and diesel *Energy Convers. Manag.*92 342–52

[36] Li J, Yu X, Xie J and Yang W 2020 Mitigation of high pressure rise rate by varying IVC timing and EGR rate in an RCCI engine with high premixed fuel ratio *Energy*192 116659

[37] Li Y, Jia M, Chang Y, Xu Z, Xu G, Liu H and Wang T 2018 Principle of determining the optimal operating parameters based on fuel properties and initial conditions for RCCI engines *Fuel*216 284–95

[38] Li Y, Jia M, Liu Y and Xie M 2013 Numerical study on the combustion and emission characteristics of a methanol/diesel reactivity controlled compression ignition (RCCI) engine *Appl. Energy*106 184–97
[39] Lim J H and Reitz R D 2017 ICEF2013-19140 1–12
[40] Ma S, Zheng Z, Liu H, Zhang Q and Yao M 2013 Experimental investigation of the effects of diesel injection strategy on gasoline/diesel dual-fuel combustion *Appl. Energy*109 202–12
[41] May I A 2018 *AN EXPERIMENTAL INVESTIGATION OF LEAN - BURN DUAL - FUEL COMBUSTION IN A HEAVY DUTY DIESEL ENGINE A thesis submitted for the degree of Doctorof Philosophy by Department of Mechanical, Aerospace and Civil Engineering Brunel University London*
[42] Nazemi M and Shahbakhti M 2016 Modeling and analysis of fuel injection parameters for combustion and performance of an RCCI engine *Appl. Energy*165 135–50
[43] Nieman D E, Dempsey A B and Reitz R D 2012 Heavy-Duty RCCI Operation Using Natural Gas and Diesel 270–85
[44] Noh H K and No S Y 2017 Effect of bioethanol on combustion and emissions in advanced CI engines:
HCCI, PPC and GCI mode – A review *Appl. Energy*208 782–802
[45] Pachiannan T, Zhong W, Rajkumar S and He Z 2019 A literature review of fuel effects on performance and emission characteristics of low-temperature combustion strategies *Appl. Energy*251 113380
[46] Pan S, Liu X, Cai K, Li X, Han W and Li B 2020 Experimental study on combustion and emission characteristics of iso-butanol/diesel and gasoline/diesel RCCI in a heavy-duty engine under low loads
*Fuel*261 116434
[47] Paykani A, Kakaee A H, Rahnama P and Reitz R D 2016 Progress and recent trends in reactivity- controlled compression ignition engines *Int. J. Engine Res.*17 481–524
[48] Paykani A, Kakaee A H, Rahnama P and Reitz R D 2015 Effects of diesel injection strategy on natural gas/diesel reactivity controlled compression ignition combustion *Energy*90 814–26
[49] Poorghasemi K, Khoshbakhti R, Ansari E, Khoshbakht B, Shahbakhti M and Naber J D 2017 Effect of diesel injection strategies on natural gas / diesel RCCI combustion characteristics in a light duty diesel engine *Appl. Energy*199 430–46
[50] Ramasankaran A P, Duraisamy G and Rangasamy M 2019 INFLUENCE OF SPLIT INJECTION MASS AND INJECTION PRESSURE ON METHANOL / DIESEL RCCI COMBUSTION IN A COMPRESSION IGNITION ENGINE 1–6
[51] Reijnders J, Boot M and de Goey P 2016 Impact of aromaticity and cetane number on the soot-NOx trade-off in conventional and low temperature combustion *Fuel*186 24–34
[52] Reitz R D 2013 Directions in internal combustion engine research *Combust. Flame*160 1–8
[53] Sarathy S M, Oßwald P, Hansen N and Kohse-Höinghaus K 2014 Alcohol combustion chemistry *Prog.*
*Energy Combust. Sci.*44 40–102
[54] Reitz R D and Duraisamy G 2015 Review of high efficiency and clean reactivity controlled compression ignition (RCCI) combustion in internal combustion engines *Prog. Energy Combust. Sci.*
[55] Sharpe B and Delgado O 2016 Engine technology pathways for heavy-duty vehicles in India 1–10

[56] Splitter D, Hanson R, Kokjohn S, Wissink M and Reitz R 2011 Injection effects in low load RCCI dual-fuel combustion *SAE Tech. Pap.*

[57] Splitter D, Reitz R and Hanson R 2010 High efficiency, low emissions RCCI combustion by use of a fuel additive *SAE Tech. Pap.*3 742–56

[58] Thangaraja J and Kannan C 2016 Effect of exhaust gas recirculation on advanced diesel combustion and alternate fuels - A review *Appl. Energy*180 169–84

[59] Tian Z F and Abraham J 2014 Application of computational fluid dynamics (CFD) in teaching internal combustion engines 42

[60] Tripathi G, Dhar A and Sadiki A 2018 Recent Advancements in After-Treatment Technology for Internal Combustion Engines—An Overview 159–79

[61] Wang H, DelVescovo D, Reitz R D and Yao M 2015 Numerical Study of RCCI and HCCI Combustion Processes Using Gasoline, Diesel, iso-Butanol and DTBP Cetane Improver *SAE Int. J. Engines*8 831–45

[62] Wu Y and Reitz R D 2015 Effects of exhaust gas recirculation and boost pressure on reactivity controlled compression ignition engine at high load operating conditions *J. Energy Resour. Technol. Trans. ASME*137 1–8

[63] Xu G, Jia M, Li Y, Chang Y and Wang T 2018 Potential of reactivity controlled compression ignition (RCCI) combustion coupled with variable valve timing (VVT) strategy for meeting Euro 6 emission regulations and high fuel efficiency in a heavy-duty diesel engine *Energy Convers. Manag.*171 683–98

[64] Dhale A A, Awari G K and Singh M P 2010 Analysis Of Internal Combustion EngineWith A New Concept Of Porous Medium Combustion For 14 943–56

A systematic review on real time exhaust gas sensing system for on board sensing of harmful gases in IC engine

S M Bante[a], S R Karale[a] and G K Awari[b]

[a] Department of Mechanical Engineering, G H Raisoni University Amaravati, India.
[b] Department of Automobile Engineering, Government Polytechnic, Nagpur, India

E-mail:shrikant.bante769@gmail.com

Abstract. Air pollution causes detrimental effects on human health and the atmosphere. The emissions associated with transport represent a major component of aerial pollution. In recent years, air pollution is becoming a very important issue in India and around the World. Conventionally, the most popular dynamometer testing has been widely used to measure vehicle exhaust emission and these measured emissions have been used as primary data for modeling transport related air quality impact. However, during such conventional emission testing, real driving conditions are not strictly considered. Nowadays, real-time vehicle emission measurement is mainly focused by the most of the researchers for analysis purposes. In the present paper, the various real-time vehicle emission measuring technologies such as vehicle emission measurement by using RFID technology, emission measurement using gas sensor technology, emission measurement using gas sensor technology with the use of IoT and emission measurement using portable emission measurement system (PEMS) is discussed.
Article I.

Keywords: Air pollution, Radio-frequency identification (RFID), Internet of things (IoT), Portable emission measurement system (PEMS).

1. Introduction

Air pollution constitutes the foremost pressing environmental health risk facing our world population. It's calculable to contribute toward seven million premature deaths a year, whereas near about ninety-two percent of the world's population is estimated to breathe toxic air (WHO, 2016). In less-developed nations, 98 percent of children under five breathe poisonous air. As a consequence, air pollution is the biggest cause of death for children under 15 years of age, killing 600,000 every year [1] [2]. The World Air Quality Report 2019 presents PM2.5 data made publicly accessible during 2019, data reported in this report are focused on real-time or roughly real-time monitoring by government sources and independently controlled and validated non-governmental air quality monitors. This report concluded that according to region wise South Asia, Southeast Asia, and Western Asia bear the highest burden of fine particulate matter (PM2.5) emissions overall, with only 6 of 355 cities included meeting WHO annual goals in these areas collectively. During 2019, the study highlights the world's top 30 most polluted cities, 21 in India, 27 in South Asia, and all the top 30 cities in Greater Asia [3]. In India day by day air pollution is becoming a serious problem from both the human perspective and environmental perspective [4]. Traffic-related emissions constitute a major component of airborne pollution [5], and it is largely responsible for climate change [6]. Transport emissions, primarily road transport, contribute significantly to the number of greenhouse gases in the environment, so the amount of exhaust gas emissions produced by motor vehicles needs to be accurately measured . Measuring the emissions of exhaust pollutants from car engines is very important because of their adverse environmental impact [7]. In urban areas the high inflow of vehicles decreasing air quality and causing more air pollution that leads to serve health diseases [8]. After decades of study and laboratory studies on livestock, the US Environmental Protection Agency (EPA) report concluded that the contaminants are carcinogenic and thus contribute significantly to the production of lung cancer; carmakers treat the issue of exhaust pollution with priority [9]. There is need to develop a system for measuring the emissions of exhaust pollutants from vehicle which will be simple to use and does not take a long time to learn, and that can display results in the form of parameters that people can easily understand[10]. Studies on automotive exhaust emissions under real operating conditions have clearly demonstrated that the level of natural vehicle emissions is higher than the level of laboratory test emissions [11] [12]. The Real Driving Emission (RDE) test concluded that the emission of nitrogen oxides is greater than that obtained in the laboratory test [13]. They, therefore, suggest using the test results of the RDE (Real Driving Emission) to develop the engine control system. Emissions from diesel vehicles are responsible for more air pollution related death worldwide [14] [15]. In the majority of developing countries, particularly India, the impact of diesel engine emissions is growing. In 2016, the Indian government decided to fully bypass the Bharat stage (BS) V standards and to comply with increasingly strict BSVI standards by 2020, which will reduce particulate matter (PM) and nitrogen oxide (NOx) emissions. [16]. In the BS-6 diesel engine to reduce the level of NOx and particulate matter, all vehicle manufacturers have been mainly focused on emission control devices to protect the environment from vehicular emission. One of the emission control devices used in the BS-6 diesel engine is called a "Diesel Particulate Filter (DPF)" which fixes and filter harmful carbon particles emitted during the combustion cycle in the exhaust system, preventing the release of these particles into the atmosphere. The DPF filter is designed to reduce diesel particulate matter and soot emissions up to 80% but this device also having some limited capacity after that it may be blocked so there is a need to regenerate the DPF [17-18-19].The contribution of exhaust emissions from vehicles to the local atmosphere is many times greater than the emissions predicted by traditional emission measurement methods. The national clean air programme (NCAP Ministry of Environment, Forest & climate change , Government of India 2019) report 2019 [20] concluded that there is a need to develop techniques and methods for measuring harmful emissions in real time due to the extreme risk posed to human health and the environment. Therefore the real time sensing system must be designed to monitor air pollution through the transport sector.

2. Factors affecting vehicular emissions

Fuel intake and exhaust emission rates are closely related to engine load and vehicle speed [21]. Factors that directly affect engine load are vehicle acceleration, engine friction, wind resistance, roadway efficiency, tire-roadway friction and accessories such as air conditioning. In turn, these engine load sources are calculated by a combination of vehicle characteristics such as motor displacement, vehicle mass, transmission performance, etc. and operating conditions of vehicles. Operating conditions include vehicle speed and acceleration are influenced by traffic congestion, driver attitude, traffic signals, posted speed limits, etc. [22]. In the summer of 2016, researchers (Granié, M.A. et.al. 2011; Holmén, B.A.et.al.1998) performed an on-road emissions measurement experiment in the city of Logan, UT, USA to determine the effects of driver experience and gender and driving behaviour on vehicle emissions, where the driving of the same vehicle along the same route was measured in four different driver groups: Experienced versus Inexperienced, female versus male. The results revealed that experienced drivers and male drivers released more HC and NOx than the control driver, which was not the case with both young and female drivers [23] [24]. Jawad H. Al-rifai[25], found in his research that road grades, speed and type of vehicle had a substantial effect on gas emissions rates. Since there is no uncertainty about the fact that heavy traffic on road is more responsible for generating more pollution hence significant attention is needed to evaluate and quantify the factors which are responsible for these emissions. The following listed factors are directly or indirectly affect the emission level [26].

IOP Conf. Series: Materials Science and Engineering 1170 (2021) 012012 doi:10.1088/1757-899X/1170/1/012012

Type of vehicle: Vehicles in this category can be categorized into the following categories, such as vehicle brand, vehicle model, engine size, vehicle era, vehicle technology, vehicle maintenance and categories of exhaust systems, vehicle tire pressure.
Type of Fuel: Exhaust emission emitted from vehicles depends upon grade, type of fuel such as CNG, biodiesel, petrol, diesel.
External Conditions: External conditions mainly include environmental conditions such as pressure, temperature, and humidity.
Driving activity: Driving activity includes normal driving, aggressive driving, cruising, idling, accelerating, and decelerating.
The vehicular emissions also rely upon proper contact between driver and vehicle. The vehicular emissions can be minimized by identifying user optimal behavior while driving and intimate the drivers about the problems associated with the inefficient style of driving, as well as informing them about its possible solutions.

3. Implementation of Emission Norms in India

In India, as proposed by the Auto Fuel Policy and the productive development of new technologies, the Indian automotive industry is primarily working towards regulatory emissions. As Indian safety requirements are compliant with Global Technical Regulations (GTR) and UN Regulations, vehicle technology in India is now on par with international benchmarks [27]. The first concept of emission standards in India was introduced for petrol engines in 1991 and diesel engines in 1992. From April 1995, in the four metro cities Delhi, Kolkata, Chennai, and Mumbai, new petrol passenger cars with the mandatory fitting of catalytic converters were launched along with the availability of unleaded petrol [28]. The Bharat Stage-I (BS-I) emission standards for passenger cars and commercial vehicles were implemented in India in 2000 to meet the Euro I standard. In Delhi, Mumbai, Chennai and Kolkata, further Euro II equivalent Bharat Stage II standards were in place from 2001 onwards. In August 2002, the first auto fuel policy was implemented, in which four-wheeled vehicles have followed the BS III norm. In April 2010, Bharat Stage IV was introduced for 13 metro cities, and the rest of the country switched to Bharat Stage III. The criteria for Bharat Stage IV were expanded to an additional 20 cities in October 2014. India's National Capital Region (NCR) was in extreme condition due to a massive rise in levels of air pollution. This drastic situation has led the Indian Government to take the strongest decision to bypass the BS V emission standard, which was introduced in 2020, and to advance the introduction of the BS-VI emission standard from 2024 to 2020[29].

Table.1: History of emission standards implementation in India

Emission Standard	Reference	Year of Introduction	Implemented Region
Bharat Stage –I(BS-I)	**EURO 1**	2000	Nationwide
		2001	Kolkata Chennai Mumbai NCR*
Bharat Stage –II(BS-II)	**EURO 2**	2003	11 cities# NCR*,
		2005	Nationwide
		2005(04)	11 cities#,NCR*
Bharat Stage –III(BS-III)	**EURO 3**	2010	Nationwide

		2010(04)	NCR*, 13 cities
	EURO 4	2015(01)	North India Rajasthan Border Districts(9 States)
		2015(07)	Near about plus 29 cities in the stateof Haryana, Uttar Pradesh, Rajasthan, and Maharashtra
Bharat Stage –IV(BS-IV)		2016(04)	East India, parts of western and south India.
		2017(04)	Nationwide
Bharat Stage –V	**EURO 5**	Skipped	Skipped Nationwide
Bharat Stage BSVI (BS-VI)	**EURO 6**	2020(04)	Nationwide

Source : Elavarasan.G. et.al.2019 [30]

4. Role of On-Board Diagnostic Device (OBD)

The major automotive manufacturers have fitted their cars with an onboard diagnostic system (OBD) to monitor vehicle exhaust emission emissions according to the norm [31]. The system, which has stringent emission significance and its essence in the mid – 90s is vehicular emissions, is used [32]. The on-board diagnostic is an intelligent device in which different types of sensors are embedded into the important areas of the vehicle that track the output and report the device whether those areas are working according to the specified parameters of the vehicle or not. The sensors activate a warning light on the vehicle's dashboard when any variations are detected, indicating that the area needs to be repaired. It is a computer-based system that detects failures associated with exhaust emissions in light-duty trucks, passenger cars, and even in heavy-duty vehicles from some years onwards. Although OBD-II diagnoses any emission-related fault, it does not guarantee that the driver understands the alarm of the MIL (Malfunction Indicator Lamp) fault indicator and fixes the failure of the vehicle promptly [33]. The OBD-III is developed as a new generation of OBD systems, featuring wireless transmission of fault information. Effective exhaust emissions control and cleaning systems are designed by original equipment manufacturers (OEMs) to meet the legal specifications of emission standards. A so called onboard diagnostic system must supervise these systems and the associated components. In 1990, OBD was first used in California to track components and parameters for pollution control. The California OBD framework specifications used in light and heavy-duty vehicles have been implemented in two stages:

OBD I: The California Air Resources Board (CARB) required all new vehicles sold in 1991 in California and newer vehicles to have some basic OBD capacity to monitor some of the components of pollution control. The OBD I was relatively simple and limited to only checking some of the emission control elements, and it was not calibrated to an exact emission performance level[34] [35].

OBD II: On-board diagnostics (OBD-II) second generation vehicle diagnosis norm that was developed for diagnostics and pollution control. This offers an opportunity to create new software that provides diagnostic services for car owners with more knowledge. Improved vehicle repair and diagnostic functions and more comfortable monitoring of equipment can be provided [36].

5. Related Work

5.1 Literature Survey Based on RFID Technology

Various emission control systems have been explored in various kinds of literature over the year. Chi-man vong et.al.2014 [37], in the present paper author mainly focused on the emission control system for developing countries is discussed. The systems are made up of RFID tags to which the lambda sensor is attached with an analogue to digital converter. The lambda sensor mounted at the exhaust pipe of the vehicle for calculating air ratio. If the air ratio is less than one then the emission of carbon monoxide and the hydrocarbons will be increased and when the air ratio more than one nitrogen oxide will be emitted. The RFID reader is used to read the value. This RFID tag passes data to the Second RFID device. The second RFID device is attached to the 3G card from which data is sent to the database server where all information is successfully modified and saved. If the actual data does not match with the standard data, then the message is generated and sent to the vehicle owner.
Anita Kulkarni et.al.2015 [38], the onboard vehicle (OBU) device is installed, which is used to collect vehicle exhaust data and send it to the roadside unit (RSU) installed at the traffic signal point. Received data along with traffic threshold transfer to the server side unit (SSU). The author concluded that using this method would minimise CO (carbon monoxide) exhaust emission. This system stops the vehicle and generates the message and sends it to the unique contact number stored in the GSM module if vehicle emission crosses pre-set standards.
S.P.Bangal et.al. [39], this paper's primary objective is to control road emissions and track a vehicle that causes pollution. During this process, Radio Frequency Identification (RFID) technology as an affordable and mature wireless communication technique is adopted to gather and transmit emissions data of vehicles and the Internet of Things (IoT) concept is proposed. Active RFID and active RFID tags are designed to gather emission data from the exhaust system of the car. When vehicles equipped with the RFID tag drive inspect the range, the RFID reader can receive emission data. ADC data collected will be sent to the ARM Microcontroller and then to the RFID tag via GPRS and Cloud Web Server to the Cloud Data Base Centre and the GPRS will be located in a remote location.
Ramgiri R. et.al. [40], in this proposed system author, used a wireless sensor along with active RFID monitor vehicular pollution based on IoT. IoT plays a vital role in the proposed system to address vehicular pollution in real-time applications. Given a system equipped with a CO_2 gas sensor and SOx gas sensor to monitor the pollutants continuously to maintain the quality of the air. The RFID reader, wireless gas sensors are integrated along with the microcontroller at the monitor location. The given proposed system is placed on either side of the road. The proposed system of the given paper can identify the vehicle which is attached with RFID tags passes through the sensor node. The sensor measures the exhaust emission emitted by the vehicle and the RFID reader displayed the information in the monitoring system. The real-time data detected is further sent to the microcontroller to verify the concentration of pollutants from the vehicle. The microcontroller verifies the levels of the emissions pollutant of the air emitted by the vehicle. If the amount of pollutants is greater than the threshold, the vehicle sends an alert warning message to the owner of the vehicle. The information about the RFID of the vehicle and time and date of the vehicle, the levels of CO_2 and SOx, vehicle number send to authorized agencies. This data get stored in the server database for future analysis. The microcontroller monitors the levels of emission pollutants released by the vehicle. If the amount of pollutants is greater than the threshold, the vehicle sends an alert message to the owner of the vehicle.
Rajalakshmi et al [41] proposed another system, including the Arduino board with ATmega 328 and GSM modem. The system proposed in the paper Control air pollution on roads and track vehicles that cause pollution above a certain limit. Periodic monitoring of sensor values by the vehicle unit is noted and sent to the laptop via the server. Vehicle FC details, insurance details, periodic service details are viewed on the laptop in addition to the sensor values. If any rise in sensor readings, the date of any

IOP Conf. Series: Materials Science and Engineering 1170 (2021) 012012 doi:10.1088/1757-899X/1170/1/012012

insurance expires, a warning will be sent to the owner by FC Service. And even the transport service with respect to the insurance of cars.

Manna S. et al [42] this paper aims to control air pollution over a defined limit on roads and track vehicles that cause pollution. To resolve this problem, this paper proposes the Internet of Things (IoT). Here, system consists of the Wireless Sensor Network and Electrochemical Toxic Gas Sensors and the use of a tagging device for radio frequency identification (RFID) to track car emission records anywhere at any time. System deployed the RFID readers at a road intersection to test the prototype. It also allows the gas sensors deployed on the roadside to collect data. The vehicles that frequently drive by the intersection are equipped with a few RFID tags. The sensor data is read concurrently with data from RFID readers, and the levels of emissions from individual vehicles can be calculated in real-time over a span of five days. The vehicles that frequently drive by the intersection are equipped with a few RFID tags. This work presents the design of a device to provide a solution for the identification of environmental pollution-causing vehicles. This work seems promising to resolve the issue based on the experimental results obtained by the author. Compared to other non-RFID-based vehicle detection systems, the device has an essential significance in terms of cost and effectiveness.

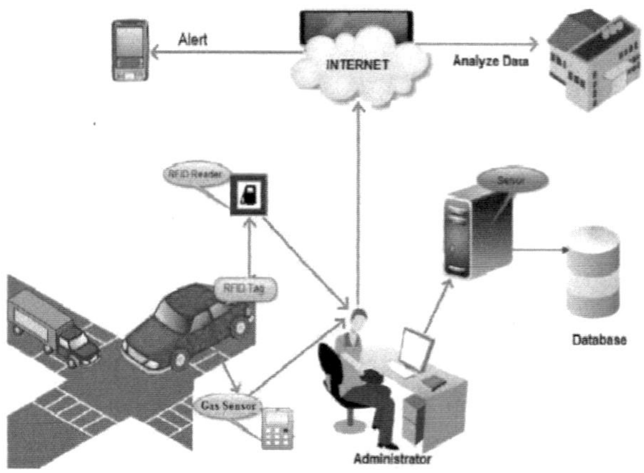

Figure 1. Proposed Emissions Control Method Using RFID technology (Source [39]: Ramgiri R. et.al.IEEE, 2015)

5.2 Literature Survey Based on Gas Sensor Technology

G Arun Francis et.al [43], In this system, the hydrocarbon, carbon monoxide, and nitrogen oxide amounts emitted from the exhaust are monitored by mq2, mq7, mq135 sensors mounted at the vehicle exhaust. The analogue value obtained from the sensors is analyzed by the Internet-connected Wi-Fi controller. The value of the sensors is displayed on LCD and cloud on an on-going basis. When the value obtained from the sensor exceeds the threshold limit, the controller alerts the user via the vehicle owner's LCD and database. Io T allows the computer to change the value of the cloud.

P Arun Mozhi et.al. [44], the proposed system used in this paper consists of a microcontroller and a sensor to measure vehicle emissions in order to alert the government to monitor the AQI and to communicate emission related information through GSM. The proposed prototype system for monitoring and detecting the level of emissions of the individual vehicle and sending information to the Central pollution control board and RTO if the vehicle crosses the threshold limit. The present system capable of detecting nitrogen oxide (NO) and carbon monoxide (CO), Carbon oxide (Cox) by using gas sensors. Due to heat dissipation from exhaust emission, the sensors may be malfunctioned during working so that they placed it in the thermal Isolation Clip. The real-time data get collected in

PIC microcontroller and data analyzed using C coding. While comparing data any deviation occurs then with the help of GSM and Information regarding GPS modules is transmitted to the nearest control station also to the driver with the help of LCD display.
Ms.S.Menaga et.al.[45], the author suggested a vehicle monitoring system (VMS) consisting of different sensors, including O_3,NO_2,CO and PM2.5.The Given system provides data on pollution, smoke, vibrations, and also GPS signals by driving a vehicle(car, bus, etc.) on the street. Using IoT, the various parameters sensed by different sensors are registered and transmitted to a cloud server. The transmitted data is related to GPS location, weather parameters, information about the vehicle, and information about air quality. Based on transmitted data, Government agencies are easily takes decisions on traffic planning and on taking some steps to minimise emissions. For communication purposes, a system called LoRa, which is a transmission protocol, is used in VMS and is regulated by the LoRa Alliance. In this system MQ-131sensor is used to detect ozone, PM 2.5/10 sensor is used to detect particulate matter in exhaust emission of vehicle, MICS-4514 sensor is used to detect carbon monoxide and nitrogen oxide. The Sensor used in this system has a simplified drive circuit and long life, high sensitivity, wide range detection capacity.
Palconit, M.G.B.et.al.[46] , the current study focuses on the statistical assessment of the emission of PUV CO_2 and its corresponding driving patterns and road grades. The CO_2 emission is calculated with respect to vehicle specific power (VSP). The information is obtained using the CO_2 NDIR (non-dispersive infrared) sensor and GPS (global positioning system) receiver, which are calibrated and tested to ensure the reliability of the reading. Two PUVs are used for five days at the San Carlos-Talamban Campus in this study in which all selected PUVs have been fuelled with gasoline. Google Earth was used for the road grade assessment. Each PUV has a built-in Wi-Fi chip microcontroller named ESP 8266 and two sensors, namely the NEO u-blox 6 Global Positioning System (GPS) receiver and a CO_2 NDIR rapid response sensor named COZIR-WR GC-0006. The Raspberry Pi microcontroller was used to process the received data, including road grade and acceleration measurement. The processed data is stored in CSV format and synchronised with a cloud server called Amazon Web Services, which uses the Kibana dashboard to show the synced files so that they can be accessed anytime and anywhere through the Internet. In order to interpret the collected data, statistical methods were used, such as collecting clustered data with theme values and standard deviations.
Gautam A. et.al [47], the proposed system in the present paper can be used in any moving vehicle which is responsible for air pollution in the atmosphere. The MQ2 gas sensor is placed at the exhaust pipe of the vehicle with the help of wire which is routed up to the raspberry pi, which is placed behind the dashboard. At the top of the dashboard, the LCD and buzzer are mounted and linked through wires with raspberry pi. In the proposed prototype of the device, the DC motor with the motor driver IC is just an analogy of vehicle movement. In the actual vehicle, a relay is connected in place of the engine, the output of which is used to ON / OFF the engine In the actual vehicle, a relay is connected in place of the engine, the output of which is used to ON / OFF the engine. All components used in this system get powered using the existing battery system of the car. In the python code, acceptable ppm levels of CO, LPG, and smoke are already fixed. The driver can easily monitor the ppm level of the gases on LCD. An alert in the form of a buzzer beep is given to the driver whenever the ppm level of the gases exceeds the threshold of appropriate ppm levels. This warning is an indicator to get your vehicle serviced in a given time frame by the driver. Negligence of the alarm allows the ppm level of gases to rise further. A challan is produced and sent on the email Id of the driver as the ppm level of gases crosses the predefined appropriate limits (threshold). The proposed system helps to maintain the level of gases within the specified limit. The results show that the device meets the requirement of an embedded sensor-based system that can screen and monitor air pollution using IoT with challan generation from anywhere in the world. The findings indicate that the full framework is checked and implemented effectively.

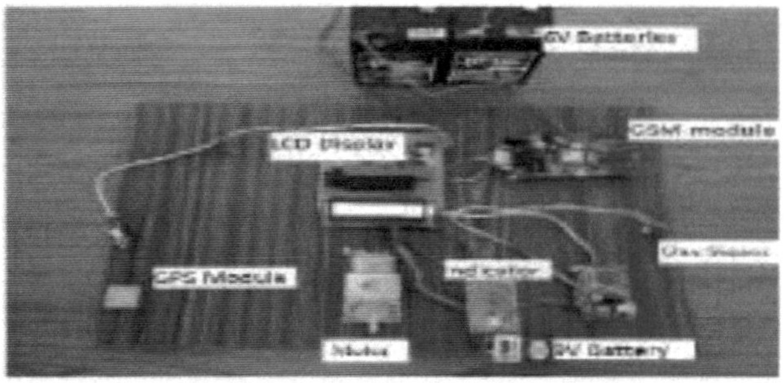

Figure2. Real-time emission sensing system Using gas sensor (Source: P Arun Mozhi et.al[41])

5.3 Literature Survey Based on PEMS (Portable emission measurement system) Technology.

Jerzy Merkisz et.al. [48], in the present paper author emphasis on real driving emission test for the light-duty vehicle (LDV), heavy-duty vehicle (HDV). The actual driving emissions test was carried out under real traffic conditions with portable emission measurement system (PEMS) equipment. PEMS such as SEMTECH DS is used to measure CO, CO_2, No, NO_2 emission from the exhaust of the vehicle. To measure the PM [mg / m3] concentration, PEMS AVL MSS (Micro Soot Sensor) is used. TSI 3090 EEPS™ determines the size distribution of PM [nm] is used during the test. The paper discussed the problem of measurement of CO, HC, NOx, PM emissions, and related methodologies. After the completion of the test, it was reported that the main problem is the emission of NOx and PM, which is normally higher than the level of emission, which also indicates that the actual driving emission method is very complex, but is the only way to obtain information on the actual emissions of road exhaust that cannot be obtained in the laboratory test.

O'Driscoll R, ApSimon et.al. [49], the new Euro 6 diesel passenger cars are used in this paper to calculate NOx and primary NO_2 road emissions using PEMs. Test route of Euro 6 diesel passenger vehicles, consisting of parts of the city and motorway. The vehicles used for the exhaust gas recirculation (EGR) mounted est, lean NOx traps (LNT), or selective catalytic reduction (SCR). Measurements and emission factors based on COPERT suggest that PEMS measurements are 1.6 times higher on average than COPERT estimates for NOx and 2.5 times higher for NO_2. The result shows that NOx emissions are 1 to 22 times the limit of form approval. The mean emission of NOx, 0.36 (sd. 0.36) g km1, is 4.5 times the limit of Euro 6 comparisons between PEMS.

Barouch Giechaskie et.al. [50], The RDE procedure has been described in this paper. This paper focuses on the test technique used by Portable Emissions Measurement Systems (PEMS) to check emissions of gaseous pollutants and particle numbers during a wide variety of typical road operating conditions. Total RDE test procedures are divided into the following steps such as Vehicle selection, 2 vehicle preparation, trip design, trip execution, trip verification, and calculation of emissions. Out of these mentioned steps, vehicle preparation and trip execution are described in a better way in this paper. This paper gave examples of trip verification and the calculations of emissions.

Wang H et.al. [51], in this paper study, by using plume chasing and PEMS simultaneously, comparative measurements of 12 HDDTs in China were performed. Multiple chasing tests were carried out on both freeways and local roads during the analysis for each vehicle. In this paper, different methods of data processing and key parameters are used to refine an algorithm to measure emission factors based on fuel consumption. The comparative analysis shows that NOX emissions could be satisfactorily estimated by the plume chasing process. The result shows that the relative errors of vehicle-specific emission factors for NOx were within ±20 percent compared to the PEMS results for all tested vehicles with multiple chasing results averaged.

variance in trip dynamics in the current study consists of extreme driving, softness in driving considered during the real driving test. The result obtained gave excellent insights into the possible modifications that may be needed for Indian conditions suitable for RDE testing. In this work, the PEMS instrument used was a Horiba OBS-One. The result shows that Route-1 emissions of NOx are higher compared to Route-2. This concludes that when traffic conditions are regulated, NOx emissions are lower and thus the selection of routes will probably affect the results of the RDE test. author finally conclude that ,Overall, under Indian conditions, real-world data should be collected and analysed in order to define the base procedure, rather than adjusting and adopting the EU RDE procedure. Multiple RDE tests should be performed and evaluated once the basic conditions for India are specified.
Varella R.et.al [52], three PEMS with laboratory analyzers attached to the tailpipe and the dilution tunnel were compared in this paper. The tests were performed on two Euro 6 vehicles (one petrol and one diesel) conducting the World Harmonized Light Vehicles Test Cycle (WLTC) and on a chassis dynamometer with a pre-recorded RDE cycle. The results showed that the variations between the PEMS gas analyzers and the laboratory references were generally less than 2% for CO_2 and 5% for NOx. With just a few exceptions, CO_2 and NOx mass emissions were less than 10 percent and 15 percent, respectively. The measurements of the exhaust flow rate were 10 percent at low speeds (urban conditions) and 5 percent at higher speeds.
Dinodia, H. et.al.[53] this paper explores the detailed experimental results of Real Driving Emission experiments evaluating after-treatment device configurations considering various conditions such as test paths, test days, driving behaviour etc.

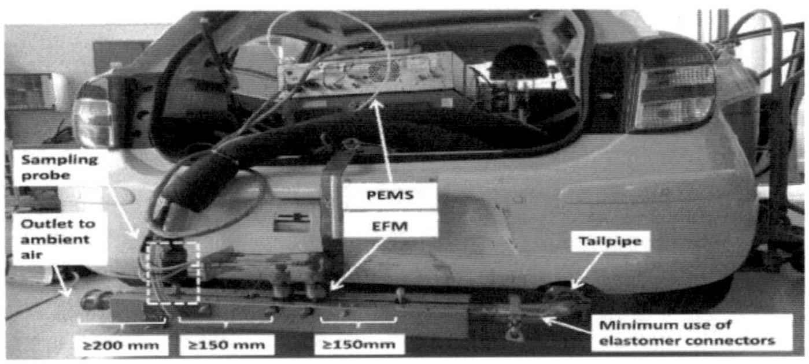

Figure 3. PEMS Installation (Source: Barouch Giechaskie et.al. [49])

Exhaust emissions emitted from vehicles get measured in real- time by various above mentioned technologies. Many of the researchers explore new technologies for measuring real-time exhaust emissions from the vehicle as given in the following table:

Table 2: Researches on real-time exhaust emission sensing system

Authors	Type of vehicle	Emission measuring device	Result & conclusion
Rohit Jaikumar et.al. (2016) [54]	Passenger Cars	AVL Digas analyzer	The findings showed that emissions in the real world were many times higher than the traditional emission factors are compared. In addition, it is ossible to use the ANN model to estimate traffic pollution under real-world conditions.

Chao-Linag Hsieh et.al. (2017)[55]	Passenger Cars	Gas sensor	In addition to obtaining vehicle information through OBD-II, the VMS suggested by this paper might also collect pollutants information via various gas sensor modules. In addition, the GPS module enables drivers to locate their cars on the street. And all of the data will be collected and viewed by users to browse the information in a graphical user interface.
Berk Demirgok.(2018)[56]	Heavy-duty (HD) vehicles	Portable Emission Measuring device (PEMS)	This research presented a detailed analysis of two different approaches for in-use emissions evaluation, namely US-NTE and EU-WBW protocols for HDD vehicles Evaluation of on-board NOx sensors along with the developed exhaust flow model is performed to calculate in-use NOx emissions within the revised NTE in-use regulatory protocol.
M Lutfie et. Al.(2018) [57]	Heavy-duty (HD) vehicles	Gas Emission sensor Mobile gas analyzer	The linear relationship between emission pollutants and vehicle speed was found in the current paper. As vehicle speed increases the high amount of pollutants released such as CO_2,NOx, HC, and CO.

Artur Jaworski et.al. (2018)[58]	Light Motor Vehicle	Portable Emission Measuring device (PEMS)	This paper provides an emission data for real driving cycles of various types of vehicles by using HORIBA OBS 2200 road analysis system. The results indicate that emissions in real-world traffic can cause vehicles to exceed emission limits. The urban and highway segments had the most exceeded pollution limits, while the rural segment had the lowest. In terms of EURO 5,The THC (exceeding by 9%) and CO (exceeding by 49%) emission limits were not included in the emission limit.

6. Conclusion

A real-time emission measurement system is needed to control air pollution caused by the transport sector. Real-time emission monitoring technologies are important research tools, as various factors affecting the exhaust emission of the vehicle are significantly considered while measurement. In the Real Driving Emissions (RDE) test, real-time sensing systems are used to measure pollutants such as NOx and CO emitted by cars when driving on the road. RDE assures that vehicles produce low emissions under on-road conditions. RDE does not eliminate the laboratory test of the WLTP (Worldwide Harmonized Light Vehicles Test Procedure). RDE is used to validate the outcomes of WLTP in real life, thus ensuring that vehicle produce low emissions of pollutants, not only in the laboratory but also on the road. Europe is the world's first region to introduce such on-road tests. With the implementation of BS6 emissions standards from 1 April 2020 and Real Driving Emissions (RDE) from 1 April 2023, the Indian automotive industry is set to witness the largest emissions reforms of the last two decades. Additionally, there are still a lot of unanswered questions regarding the RDE test protocol for India. The adoption of the European RDE test procedure, as it is, maybe a challenge for India due to the significant differences between India and Europe in terms of weather, geography, demography, road infrastructure, driving behaviour, and fuel quality. In the present paper, various technologies used for real-time exhaust emission measurements have been discussed. Many researchers are giving a decent contribution to the development of new technologies using different methods for measuring vehicular emission in real-time. This paper has briefly reviewed the available literature on the real-time measurement of vehicular exhaust emissions.

References

[1] WHO 2016 who releases country estimates on air pollution exposure and health impact. Retrieved February 17, 2020, from https://www.who.int/news-room/detail/27-09-2016-who-releases-country-estimates-on-air-pollution-exposure-and-health-impact.

[2] WHO. (2018, October 29) More than 90% of the world's children breathe toxic air every day. Retrieved February 17, 2020, from https:// www.who.int/news-room/detail/29-10-2018-more-than-90-of the world%E2%80%99s-children-breathe-toxic-air-every-day.

[3] IQAir report 2019 World air quality report Region & City PM2.5 Ranking (Accessed: 1- sept-2020).

[4] Central Pollution Control Board, Monitoring Network, 2019 http://cpcb.nic.in/monitoring-network-3/ (Accessed: 1- Aug- 2020).

[5] Karl Ropkins, Joe Beebe, Hu Li, Basil Daham, James Tate, Margaret Bell and Gordon Andrews 2009 Real-world vehicle exhaust emissions monitoring: review and critical discussion, Critical Reviews in *Environmental Science and Technology*,**39:2,** 79-152.

[6] NCAP, 2019 National clean air programme. Central Pollution Control Board. Ministry of Environmental Forests and Climate Change, The Government of India [Accessed: 1- sept-2020]

[7] Jaworski A, Mądziel M and Lejda K 2019 Creating an emission model based on portable emission measurement system for the purpose of a roundabout. *Environmental Science and Pollution Research* **26(21)** 21641-21654.

[8] Merkisz J, Lijewski P, Fuc P, Siedlecki M and Ziolkowski A 2016 Development of the methodology of exhaust emissions measurement under RDE (Real Driving Emissions) conditions for non-road mobile machinery (NRMM) Vehicles *IOP Conference Series: Materials Science and Engineering* **Vol. 148 No. 1**.

[9] Environmental Protection Agency 2002 Health assessment document for diesel engine exhaust online document: http://ofmpub.epa.gov/eims/eimscomm.getfile?p_download_id=36319. [Accessed: 2-Aug-2020].

[10] Rajagukguk J, Pratiwi R A, and Kaewnuam E 2018 Emission gas detector (EGD) for detecting vehicle exhaust based on combined gas sensors. *Journal of Physics: Conference Series*, **1120,** 012020.

[11] Pielecha J, Kurtyka K and Skobiej K 2018 The impact of vehicle dynamic parameters on the exhaust emission in RDE Tests. *Combustion Engines*.**175 (4),** 26-34.

[12] Merkisz J and Pielecha J 2016 Selected Remarks about RDE Test. *Combustion Engines*. **166(3)** 54-61.

[13] May J, Favre C, Bosteels D 2013 Emissions from Euro 3 to Euro 6 light-duty vehicles equipped with a range of emissions control technologies. *Internal Combustion Engines: Performance, Fuel Economy and Emissions*, 55–65.

[14] Resitoglui A, Altinisik K, and Keskin A 2014 The pollutant emissions from diesel-engine vehicles and exhaust after treatment system. *Clean Technologies and Environmental Policy* **17(1)** 15–27.

[15] Anderson P 2008 (*SCR* Catalysts for Low-Emitting Diesels. *In US Dept. of energy diesel engine emissions and energy reductions conference (DEER)*, Detroit.

[16] Singh G N and Bharj R S 2019 Experimental study of filtration behavior of diesel particulate filter in a diesel engine to meet BS-VI emission norms in India. *International Conference on Recent Advances in Fluid and Thermal Sciences Journal of Physics Conf. Series* **1276** 012078 Dubai. U.A.E.

[17] Bharj R S, Kumar R and Singh G N 2019 On-board post-combustion emission control strategies for diesel engine in India to meet bharat stage VI Norms. In *Advanced Engine Diagnostics* pp 105-125 Springer, Singapore.

[18] Boger T, Rose D, Tilgner I C, Heibel A K 2009 Regeneration strategies for an enhanced thermal management of oxide diesel particulate filters. *SAE International Journal of Fuels and Lubricants* **1(1):162–172.**
[19] Fasolo B, Hardy J P and Leroy K 2009 Exhaust fuel injection system for efficient DPF regenerations. *MTZ Worldwide* **70(7–8):26–32.**
[20] Ganguly T, Kurinji L, Selvaraj, Sarath K. Guttikunda 2020 National clean air programme (NCAP) for Indian cities: Review and outlook of clean air action plans. *Atmospheric Environment* **Volume 8** ISSN 2590-1621.
[21] Kean, A.J., R.A. Harley, and G.R. Kendall 2003 Effects of vehicle speed and engine load on motor vehicle emissions. *Environ. Sci. Technol.* **37(17),** 3739-3746
[22] Khader, Abdelhaleem, and Martin, Randal 2019 On the road testing of the effects of driver's experience, gender, speed, and road grade on car emissions. *Journal of the Air & Waste Management Association* **pp.1-35.**
[23] Granié M A, and Papafava E 2011Gender stereotypes associated with vehicle driving among French preadolescents and adolescents. *Transportation Research Part F: Traffic Psychology and Behaviour*, Elsevier, **14(5)** 341–353.
[24] Holmén B A and Niemeier D A 1998 Characterizing the effects of driver variability on real-world vehicle emissions, *Transpn Res-Part **D*** **3:2 117-128**.
[25] Al-rifai J H **2017** Correlation analysis of driving conditions and on-road emissions trends for vehicles. *Journal of Urban and Environmental Engineering* **11(1)** 63–72.
[26] Zhang K Yao L and LI G 2013 Factors affecting vehicular emissions and emission models. *ICTE* 2013.
[27] Society of Indian Automobile Manufacturers http://www.siamindia.com/technical-regulation (Accessed: 26-Aug- 2020).
[28] Sharma S, Jain S, Goel A, Mahtta R , Kumar R, Datt D, Kundu S, Aggarwal P, and Sharma P, 2014 Advancement of Fuel Quality and Vehicle Emissions Norms to Improve Urban Air Quality in India. *The Energy Resource Institute. Policy brief.*
[29] Sharma P, Yadav M, Yadav D and Vashist D 2018 A comparative study and analysis of emission norms adopted by developed and developing nations. *International Journal of Engineering Technology Science and Research* **Volume 5, Issue 1**.
[30] Elavarasan G, Kannan M and Karthikeyan D 2019 History of emission standards in India – A Critical Review. *International Journal of Research and Analytical Reviews*, **Volume 6, Issue 2, 28-35.**
[31] Feng G, Zhang J and Zhang W 2008 Light duty gasoline vehicle OBD parts detection system development, *Automotive Engineering*, **Vol. 30, Issue 7, 2008, pp. 602-604.**
[32] Yan S, Su W, Yang W 2009 Car remote monitoring system based on OBD II, *World Electronic Components* No. 5 **pp. 60-62.**
[33] Huang A, Xiang Z, Zhang T 2010 Design and implementation of real-time monitoring system of the vehicle information. *Vehicle Information Real- Time Monitoring System,* **Vol. 31 Issue 8 pp 1839-1847.**
[34] Bordoff F 2003 OBD what is it and what it means to you, Fact Sheet Fred's Garage, Apr 2003. http://www.fredsgarage.com/OBD.
[35] CARB, Personal Communication M. McCarthy, California Air Resources Board, Sept 2003-1.
[36] Kalmeshwar M, Prasad K 2017 Development of on-board diagnostics for car and it's integration with android mobile.*2nd IEEE International Conference on Computational Systems and Information Technology for Sustainable Solution (CSITSS)* **pp 1–6.**
[37] Vong C,Wong P, Ma Z and Wong Application of RFID technology and the maximum spanning tree algorithm for solving vehicle emissions in cities on Internet of Things, *IEEE World Forum on Internet of Things (WF-IoT)*, Seoul, Korea (South), 2014, pp. 347-352.
[38] Anita kulkarni1, T. Ravi Teja 2014 Automated system for air pollution detection and control in vehicles. *ijareeie*2014.0309061.

[39] Bangal S P, Gite P E , Ambhure S G, Gaikwad V M 2017 IoT based vehicle emissions monitoring inspection System. *International Journal of Innovative Research in Electrical, Electronics, Instrumentation and Control Engineering (IJIREEICE),* **Vol. 5, Issue 4.**

[40] Ramagiri R, Reddy C M and Reddy S 2015 Development of IoT based vehicular pollution Monitoring System. *IEEE*, **978-1-4673-7910-6/15.**

[41] Rajalakshmi A, Karthick S and Valarmathy S, 2015 Vehicular pollution and status monitoring using RFID. *IJARSET*, **vol. 2, pp 580-586**.

[42] Manna S, Bhunia S S and Mukherjee N 2014 Vehicular pollution monitoring using IoT. *IEEE International Conference on Recent Advances and Innovations in Engineering (ICRAIE-2014),* May 09-11, 2014, Jaipur, India.

[43] Francis G A, Dhinesh M,Lijo J A ,Hariprasad P, Balasubramanian K 2019 IoT based vehicle emission monitoring system. *International Journal of Innovative Technology and Exploring Engineering (IJITEE). ISSN*: 2278-3075, **Volume-8 Issue-5**S March, 2019.

[44] Devan P A M, Hussin F A, Ibrahim R, Bingi K and Nagarajapandian M 2019 IoT based vehicle emission monitoring and alerting system. *In Proceedings of the 2019 IEEE Student Conference on Research and Development (SCOReD),* Bandar Seri Iskandar, Malaysia, 15–17 October 2019; **pp. 161–165.**

[45] Ms Menaga S, Ms Paruvathavardhini J, Ms Kalaivani P and Mr Haribabu S 2019 Air quality monitoring system using vehicles based on the IoT, *International Research Journal of Engineering and Technology (IRJET), e-ISSN: 2395-0056,* **Volume: 06 Issue: 03 Mar 2019**.

[46] Palconit M G B and Nunez W A 2017 December. Co_2 emission monitoring and evaluation of public utility vehicles based on road grade and driving patterns: An Internet of Things application. *In 2017 IEEE 9th International Conference on Humanoid, Nanotechnology, Information Technology, Communication and Control, Environment and Management (HNICEM)* pp. **1-6.**

[47] Gautam A, Verma G, Qamar S, and Shekhar S 2019 Vehicle pollution monitoring, control and challan system using MQ2 sensor based on internet of things. *Wireless Pers Commun* **116, 1071–1085 (2021).**

[48] Merkisz J, Lijewski P, Fuc P, Rymaniak L and Ziolkowski A 2018 Measurement of exhaust emissions under actual operating conditions with the use of PEMS: Review of selected vehicles. *Improvement Trends for Internal Combustion Engines.*

[49] O'Driscoll R, ApSimon, H M, Oxley T, Molden N, Stettler M E J, and Thiyagarajah A 2016 A portable emissions measurement system (PEMS) study of NOx and primary NO_2 emissions from Euro 6 diesel passenger cars and comparison with COPERT emission factors .*Atmospheric Environment* **145, 81–91.**

[50] Giechaskiel B, Vlachos T, Riccobono F, Forni F, Colombo R, Montigny F Weiss M 2016 Implementation of portable emissions measurement systems (PEMS) for the real-driving emissions(RDE) regulation in Europe. *Journal of Visualized Experiments.*2016. 10.3791/54753.

[51] Wang H, Wu Y, Zhang K M, Zhang S, Baldauf R W, Snow R, Hao J. 2020 Evaluating mobile monitoring of on-road emission factors by comparing concurrent PEMS measurements. *Science of the Total Environment,* 139507.

[52] Varella R, Giechaskiel B, Sousa L and Duarte G. 2018 Comparison of portable emissions measurement systems (PEMS) with laboratory grade equipment. *Applied Sciences* **8(9), 1633**.

[53] Dinodia H, Mahajan N, Dwivedi V, and Khanna V 2018 Study on real driving emissions for India - an experimental approach. *SAE Technical Paper* 2018-01-0339, 2018.

[54] Jaikumar R, Shiva Nagendra S M and Sivanandan R 2016 Modeling of real time exhaust emissions of passenger cars under heterogeneous traffic conditions. *Atmospheric Pollution Research* **Volume 8, Issue 1**, 2017, 80-88, ISSN 1309-1042.

[55] Hsieh C L, Ye Z W, Huang C K, Lee Y C, Sun C H, Wen T H, Juang J Y, and Jiang J A 2017 A vehicle monitoring system based on the LoRa technique. *World Academy of Science, Engineering and Technology International Journal of Transport and Vehicle Engineering* 11, 5.

[56] B Demirgok 2019 Development of an emissions monitoring methodology using onboard NOx sensors and revision to current in-use emissions regulatory protocols, *Morgantown: WVU*, 2018.
[57] Lutfie M, Samang L, Adisasmita S A and Ramli M I 2018 Speed correlation and emission of truck vehicles on dynamic conditions. *IOP Conference Series: Earth and Environmental Science,* **140, 012112.**
[58] Jaworski A, Lejda K, Madziel M, and Ustrzycki A 2018 Assessment of the emission of harmful car exhaust com-ponents in real traffic conditions. *IOP Conf. Series: Materials Science and Engineering* 2018, 4**21, 042031.**

IOP Conf. Series: Materials Science and Engineering 1170 (2021) 012013 doi:10.1088/1757-899X/1170/1/012013

Improving Accuracy of Classification of Emotions Using EEG Signal and Adaptive PSO

Prof. Payal Ghutke[1] , Prof. Sonali Joshi[2] , Ruchita Timande[3]

[1]Ph.D research scholar in G H Raisoni University, Amravati & working as Assistant Professor in Department ofElectronics Engineering, G H Raisoni College of Engineering, Nagpur.
[2]Assistant Professor in Department of Electronics Engineering, G H Raisoni College of Engineering, Nagpur.
[3] M-Tech Research Scholar of VLSI in Department of Electronics Engineering, G H Raisoni College of Engineering,Nagpur.

a) payal.ghutke@raisoni.net
b) sonali.joshi@raisoni.net
c) ruchitamtimande@gmail.com

Abstract: For detecting the feelings or emotions of the human being by using brain signals and its classification has been attempt by many researchers. The EEG headset is used for collecting the brain signal of the subject. Because of lots of noise in the input signal taken by EEG headset, various features need to be used as a single feature that cannot give accurate output. The Number of feature used is the key for identifying the emotion of a person automatically. So, we identify various features using an AI based scheme from EEG recorded signals. This various features are saved in the database. Features include mean, maximum, minimum, std. deviation, variance, corr. Coefficient, cov. Coefficient, Median, Kurtosis, Energy, Zero crossing rate. By using Maximum Relevance Minimum Redundancy (MRMR), as per the name we arrange the features to minimum-relevance and maximum-importance of every feature. For removing essential segments PCA is used to diminish the produce feature. The proposed system will outperform and improve the accuracy of emotion detection by using brain wave and Adaptive PSO.

1. INTRODUCTION

EEG is the procedure of electrophysiological checking, this records the electrical developments of mind. EEG is noninvasive technique with terminals set close to the human scalp. In Electrocortocography obtrusive terminals are been used in which the sign from the mind that is electrical signal[1]. This electrical action going on in the scalp is recorded by the EEG headset. EEG headset is set on the scalp of the human from which the distinctive sign is recorded from the scalp of the human for feelings and emotions which is used for distinguishing or classification. EEG may moreover be valuable for diagnosing or treating the disperses like brain disability cerebrum hurt from brain injury, mind tumor that can have a collection of causes encephalopathy, rest issue, stroke. EEG can moreover be used in start to finish thought contraptions for cerebrum feature.

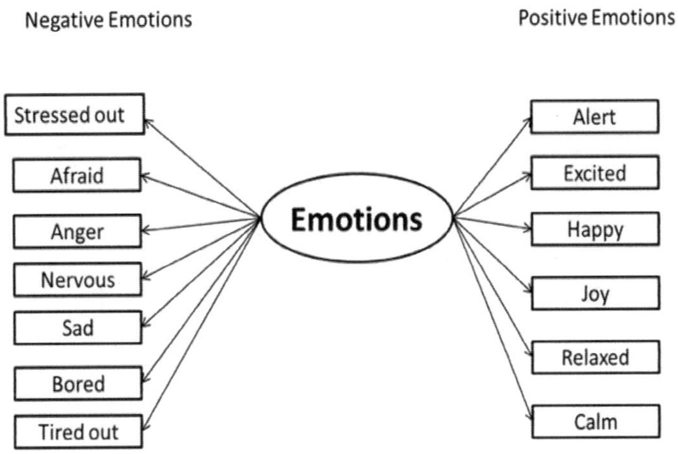

Figure 1. Types of Emotions

Lots of inefficiency come with the existing work to make it efficient new technologies can be created. Like the SVM requires a legitimate determination of part so as to perform precise location of feelings from the EEG sets[2]. Therefore, so as to tune the exactness, we are planning to coordinate a PSO and SVM based calculation which will perform characterization, and will gain from the output themselves so as to get the best possible after effects of feeling classification. The issue recognized by us is the issue of low precision given by the framework, and how to improve it utilizing the idea of adaptive swarm optimization, and its self-adaptive learning calculation.

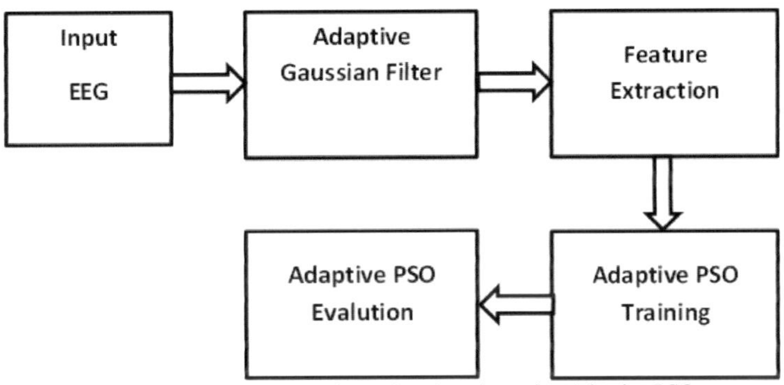

Figure 2. Block Diagram of Emotion detection using Adaptive PSO.

The above Figure 2 shows the emotion detection using Adaptive PSO[3]. In this system the different modules explained as follows:

1. Module for collecting the data: This module is use to gather information from EEG headset for all kind of feelings such as excited, sad, neutral and happy.
2. Preprocessing of Signal: Signal is preprocessed by utilizing adaptive Gaussian filter for eliminating the noise from the signal from EEG headset and to get the spike free signal. This is done for smoothening of the signal by using the filter before using classification algorithm.
3. Extraction of Feature from the Signal: With the help of statistical and wavelet features will be extracted once preprocessing is done the feature extraction in done to describe the signal as per required[4].
4. Classification of Signal: Classification of emotion can be done by using adaptive PSO which is the combination of SVM (Support Vector Machine) and PSO (Partical swarm optimization). The extracted features are further given for training and evaluation process for obtaining the classification of the emotions [3].

2. METHODOLOGY

2.1. Training

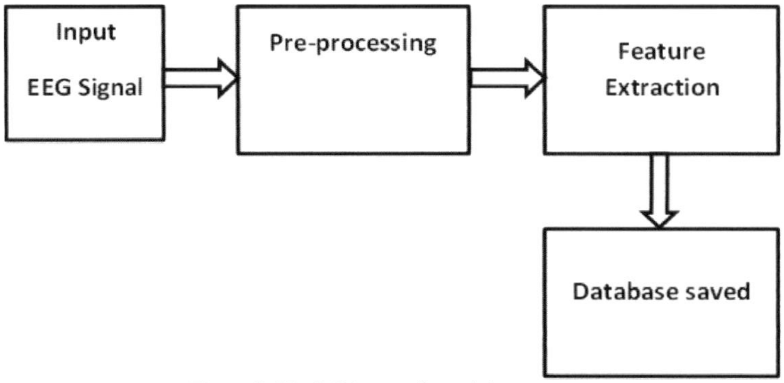

Figure 3. Block Diagram for training process.

1. Various input EEG signals of feelings & emotions such as happy, sad, neutral and excited are taken from the EEG headset for training process
2. Pre-processing using Adaptive Gaussian filter. The collected data will be pre-processed in order to obtain the noise free signals from the system. Adaptive Gaussian filter is used to remove power line interference in EEG signals. PLI noise is also known as spike noise. This noise appears because of the electrical gadgets which are operating nearby.
3. Features like Mean, Max, Min, Std. Deviation, variance, corr. Coefficient, cov. Coefficient, Median, Kurtosis, Energy, ZCR. These features are extracted using wavelet and statistical features.
4. The collected signals and their features are taken with the emotions and signals of all emotions are saved in the database with the features.

2.2 Testing

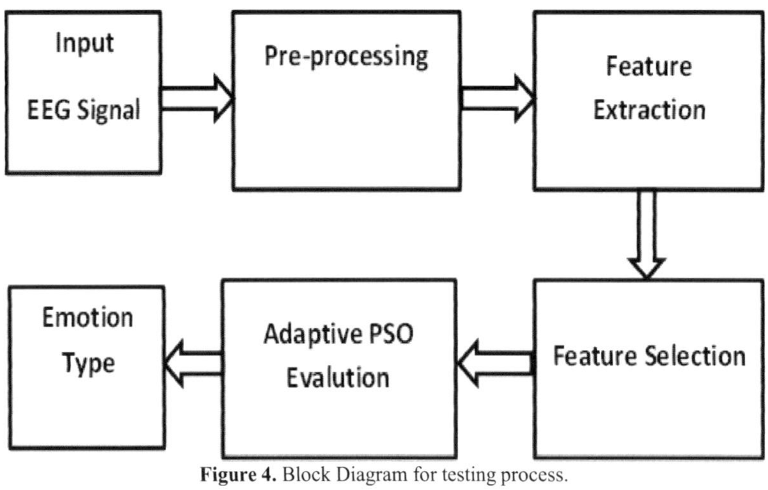

Figure 4. Block Diagram for testing process.

IOP Conf. Series: Materials Science and Engineering 1170 (2021) 012013 doi:10.1088/1757-899X/1170/1/012013

1. As the signal is taken from the EEG headset for testing the signal, this signal is pre-process by using Adaptive Gaussian filter for removing spike noise from the signal for testing the state of emotion.
2. After preprocessing the signal has been smoothen due to the filter used. After extraction of the features the features which are nearby to the original signal is consider from the database and the features which are not nearby are eliminated. Maximum features matching the original signal feature will be selected from the database and accordingly the signals will the selected
3. By using PSO signal is classified in which the original signal is the pbest i.e. personal best value and data stored (signal stored in the database) is the gbest signal i.e. group best value. pbest signal search for the nearby value of signal in the gbest signals. As the nearby value is found it display the emotion of that gbest signal which is nearer to the original signal.

3. TOOLS

The tools as well as platforms which are required to complete this project are listed below:

3.1 Project Requirements

- Hardware Components:-
  EEG Headset
- Software Requirements:-
  MATLAB
- Languages to be used:-
  Programming language C

3.2 Identification of device

Electroencephalography (EEG):

A test which is used for identification of electrical signals or action of mind by using electroencephalography (EEG Anodes (slight wires) and small plates made of metal are put on the scalp of the person and then to record the outcomes impart signals to a computer. An electrophysiological identifying technique is used for recording electric movement of cerebrum. EEG identifies covert process which is handling that doesn't require a reaction. EEG is utilized in subjects which can be unequipped for making an engine reaction. Parts of ERP can recognized in any event when the subject isn't taking care of improvements. Dislike many methods for examining response time. Phases of preparing (as opposed to simply the last final product) can be explained by ERPs. EEG is a powerful asset for following mind changes during various periods of life. EEG rest investigation can demonstrate noteworthy parts of the planning of mental health, including assessing immature mind development. In EEG, there is a superior comprehension of which sign is estimated when compared with other research procedure [5].

Adaptive Gaussian filter:

Equations Gaussian separating has been seriously concentrated in image processing and PC vision. Using a Gaussian channel for eliminating spike noise. Gaussian filter smoothen the signal by removing the noise and the signal is distorted. The utilization of a Gaussian channel as pre-processing for edge detection will likewise offer ascent to edge position removal, edges disappearing, and apparition edges. Here, the creators first audit different strategies for these issues. They at that point propose a versatile Gaussian sifting calculation wherein the channel change is adjusted to both noise characteristic and the local variance of signal [6]. Adaptive Gaussian filter is used inthe project for eliminating the noise. Noise such as PLI (Power Line Interference) or spike noise. This noise is because of the electrical equipment that is active nearby is known as PLI noise. Cancellation of this spike noise can

IOP Conf. Series: Materials Science and Engineering 1170 (2021) 012013 doi:10.1088/1757-899X/1170/1/012013

be done by using this adaptive Gaussian filter. After preprocessing the signal is send to extract the features of the EEG signal.

Feature Extraction:

Once the signal is preprocessed, then with the help of wavelet and statistical features, the features will be extracted and the signals will be described as per requirement. After preprocessing the signal has been smoothen dueto the filter used. The features are extracted from the signal [7]. After extraction of the features, the features which are nearby to the original signal is consider from the database and the features which are not nearby are eliminated. Maximum features matching the original signal feature will be selected from the database and the signals will the selected accordingly. There are many features which are extracted from the EEG signal the features can be extractedare the wavelet and statistical features. Various feature such as Mean, min., max., std. deviation, Variance, corr. Coefficient, cov. Coefficient, median, kurtosis, energy, ZCR (zero crossing rate) are calculated.

Adaptive particle swarm optimization technique:

An intelligent technique is introduced for finding a micro seismic source dependent on Particle swarm optimization (PSO) idea [8]. It kills micro seismic source finding mistakes brought about by the inaccurate velocity model. The strategy utilizes, as objective of PSO, a worldwide least of the total of squared discrepancies in between contrasts of displayed appearance time and difference of estimated appearance times[8][9]. The discrepancies are determined for all sets of indicators of seismic monitoring method. At that point, the Adaptive PSO calculation is done to find the micro-seismic source and get optimal value of the Pwave velocity[10]. The PSO calculation inertia weight, accelerating constants, the maximum flight velocity of particles, and different parameters to keep away fromPSO calculation catching by local optima at the time of solution procedure. The starting time of micro seismic occasion is evaluated by limiting the addition of squared discrepancies between demonstrated received times and thecalculate received times. This addition is determined utilizing the obtained result of micro seismic source directions and Pwave speed. Effectiveness of the PSO calculation was checked through inversion of a theoretical model and to analyze the actual information from mine blasts in many locations. The PSO calculation shows quicker, higher accuracy and convergence of micro seismic source area. In addition, there is no compelling reason to quantify the micro seismic wave velocity ahead of time. The PSO algorithm takes out adverse impacts brought by error in Pwavevelocity when finding a micro seismic source using traditional strategies [11].

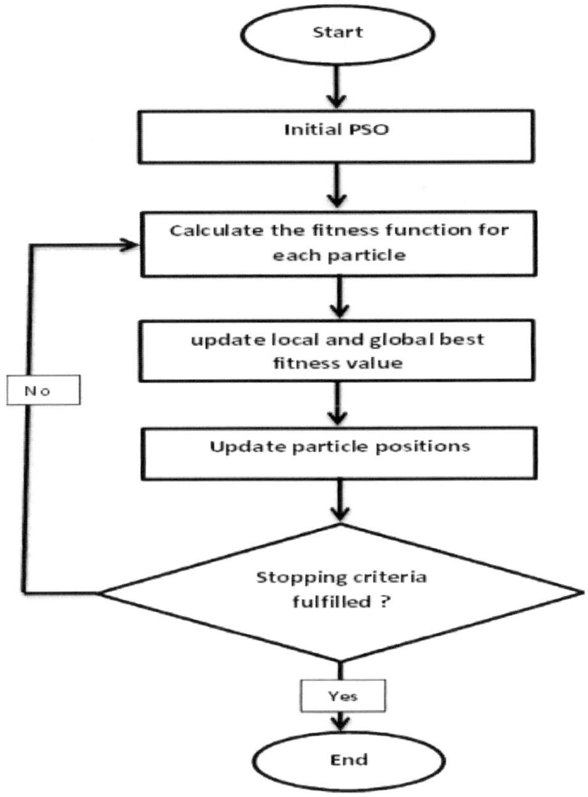

Figure 5. Flow chart of PSO classifier.

The above Figure 5 shows the flow chart of PSO classifier which explains the following steps:
1. Random particle vectors are initialized.
2. Values of fitness for each particle is found.
3. For every particle vector pbest value is calculated according to the fitness-values.
4. Then from those pbest or personal best values gbest value is obtained.
5. Update each particle's position.
6. If the target is not reached go to the step 2 and repeat the process.
7. If criteria fulfilled then end the process [12].

4. RESULT

4.1. Output Using KNN:

k-Nearest Neighbor Classifier

K-Nearest Neighbor classifier is a non-parametric methodology, which classifies a given information point as indicated by most of its neighbors. The KNN calculation finishes its execution in two stages, first finding the number of closest neighbors and second grouping the information point into specific class using first step. To find the neighbor, it utilizes Euclidean distance as given in below equation.

$$Distance(x, y) = \sqrt{(xi - yi)2}$$
$$i$$

It picks closest k samples from the preparation set, at that point it takes majority vote of their class, where k must be an odd number to avoid uncertainty.

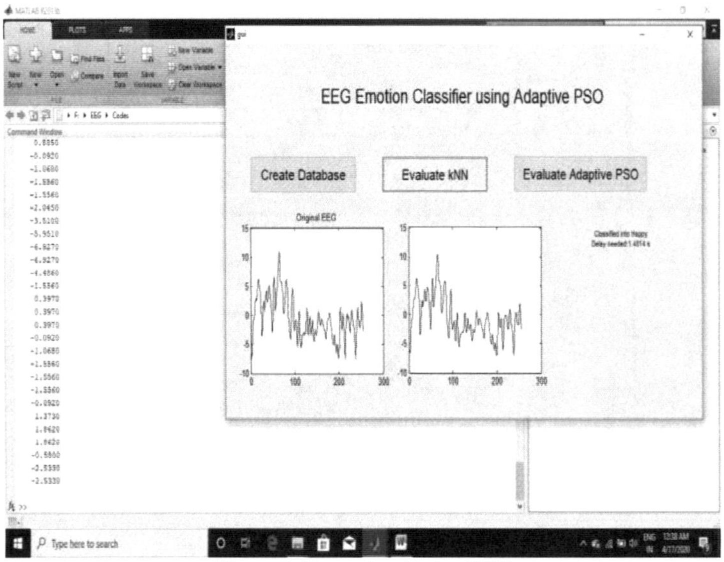

Figure 6. Happy signal using KNN

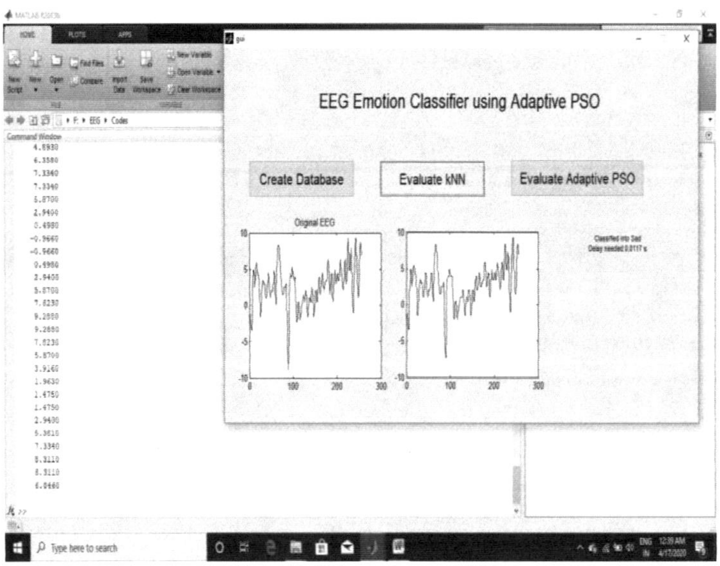

Figure 7. Sad signal using KNN

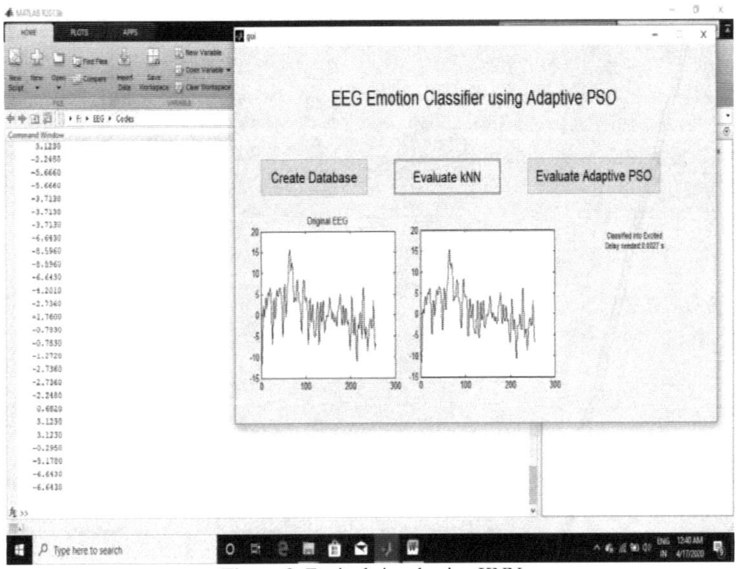

Figure 8. Excited signal using KNN

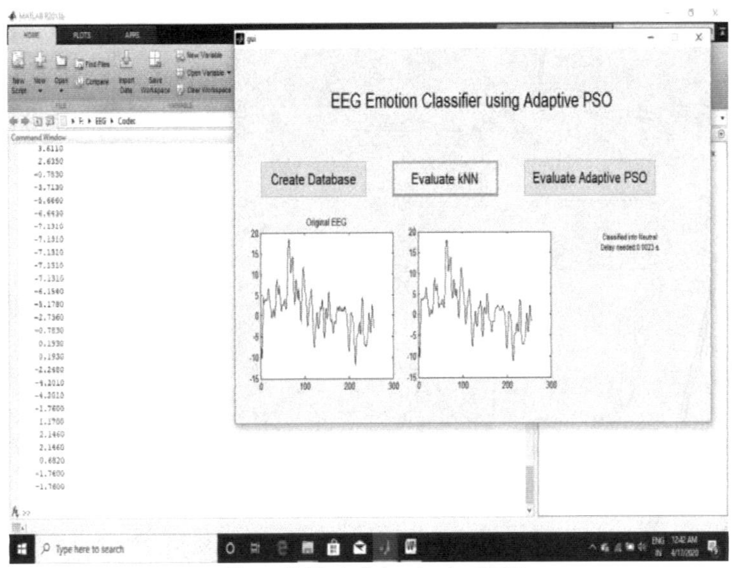

Figure 9. Neutral signal using KNN

4.2 Output using Adaptive PSO:

Adaptive PSO:-

Adaptive PSO signal is used for classification on signals. In which the original signal is the pbest and data stored is the gbest signal. pbest signal search for the nearest value of signal in the gbest signals, as the nearestvalue is found it displays the emotion of that gbest signal which is nearer to the original signal.

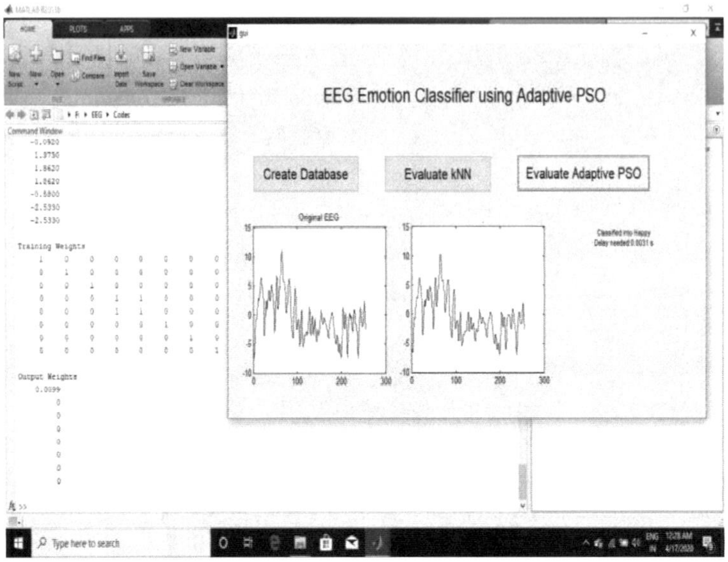

Figure 10. Happy signal using Adaptive PSO

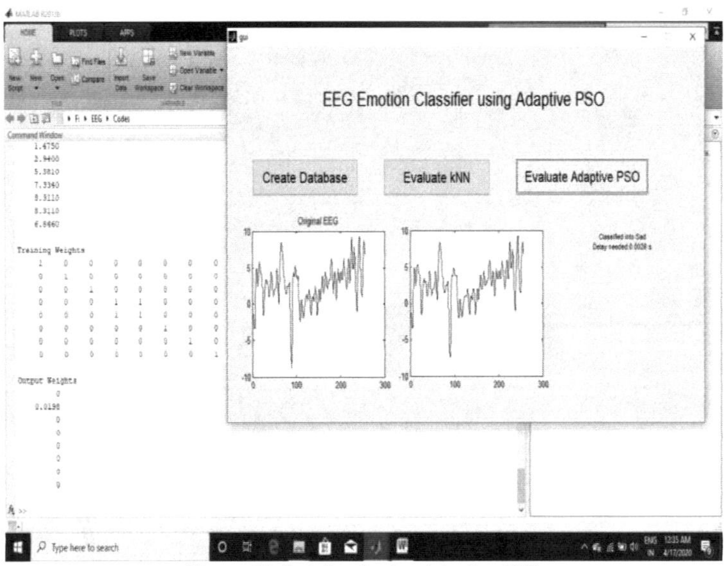

Figure 11. Sad signal using Adaptive PSO

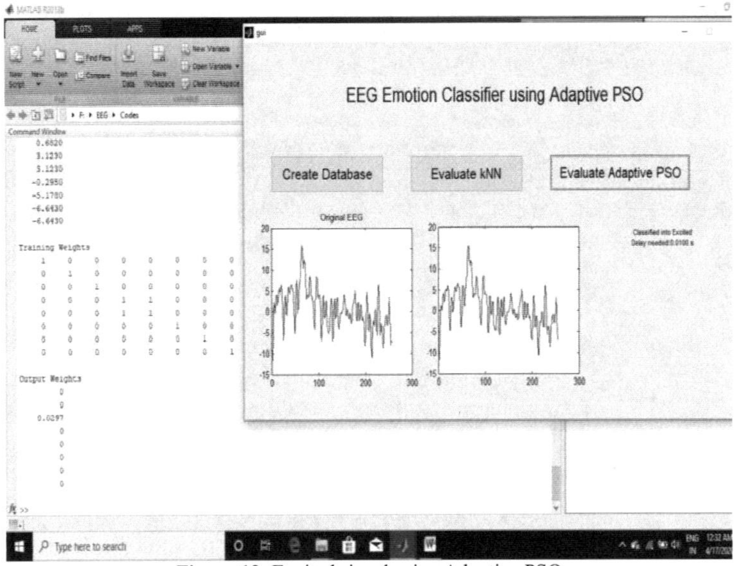

Figure 12. Excited signal using Adaptive PSO

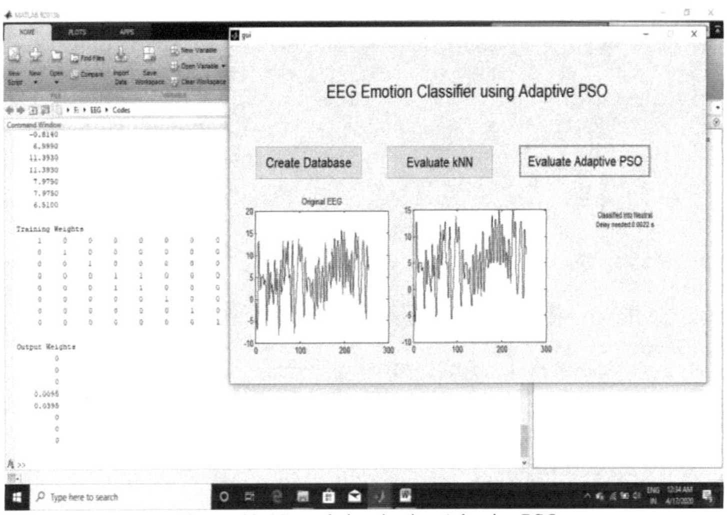

Figure 13. Neutral signal using Adaptive PSO

Table 1. Accuracy of KNN and Adaptive PSO.

Classifier	**No. Of Inputs**	**No. Of Correctly Classified**	**Accuracy**
KNN	20	17	85%
Adaptive PSO	20	19	95%

We have taken 20 inputs from EEG headset for testing and classifying the emotions of human being. By using KNN we got 17 correct output classified out of 20 and by using Adaptive PSO we got 19 correctly classifiedoutputs. So the accuracy of the classifier KNN is 85% and Adaptive PSO accuracy is 95%.

5. CONCLUSION

The proposed work have compared the output of the KNN and Adaptive PSO and calculated the accuracy of the system by using both classification techniques. We found better correctly classified output by using Adaptive PSO. Thus we have improved the accuracy of the system using adaptive particle swarm optimization technique.

REFERENCES

[1]. Niedermeyer E.; Da Silva F.L., Lippincott Williams & Wilkins, "Electroencephalography: Basic Principles, Clinical Applications, and Related Fields", (2004), ISBN 978-0-7817-5126-1.
[2]. Raja Majid Mehmood; Hyo Jong Lee, "Emotion classification of EEG brain signal using SVM and KNN", IEEE International Conference on multimedia & Expo Workshops (ICMEW), (2015).
[3]. Ruchita Timande, Payal Ghutke, "Improving Accuracy of Emotion Detection using Brain Waves and Adaptive Swarm Intelligence", International Journal of Innovative Technology and Exploring Engineering (IJITEE) ISSN: 2278-3075, Volume-9 Issue-6, (April 2020).
[4]. Hanieh Zamanian, Hassan Farsi, "A New feature extraction method to improve Emotion Detection using EEG signals.", Electronic letters on computer vision and Image Analysis 17(1): 29-44; (2018).
[5]. Electroencephalogram by city Xray Clinic, https://www.apsense.com/article/electroencephalogram.html
[6]. G. Deng and L. W. Cahill, "An Adaptive Gaussian Filter For Noise Reduction and Edge Detection", IEEE conference record nuclear science symposium and medical imaging conference, (1993).
[7]. Nitesh, Manjula B. M., "Feature Extraction of Ballistocardiogram Signal.", IJSDR, (July 2016).
[8]. Hong-Mei Sun, Jian-Zhi Yu1, Xing-Li Zhang, Bing-Guo Wang, Rui-Sheng Jia1, "Technique for solving for microseismic source location parameters based on adaptive particle swarm optimization", Manuscript under review for journal Nonlin. Processes Geophys, (2 April 2019).
[9]. Nan Li, Enyuan Wang, "Anonlinear microseismic source location method based on simplex method and its residual analysis.", Arabian Journal of Geosciences, (Nov 2013).
[10]. Hong-Mei Sun, Jian-Zhi Yu1, Xing-Li Zhang, Bin-Guo Wang, and Rui-Sheng Jia, "The adaptive particle swarm optimization technique for solving microseismic source location parameters", Nonlin. Processes Geophys., (15 July 2019).
[11]. Denisse Pasten, Rodrigo Estay, "Multifractal analysis in mining microseismicity and its application to seismic hazard in mine", International Journal of Rock Mechanics and mining Sciences, (April 2015).
[12]. Neelam Singh, Dr. Anjali Potnis, "Optimal Design of a Linear Phase FIR Filter Based on Particle Swarm Optimisation", International Journal of Advance Engineering and Research Development (IJAERD) Volume4, Issue8, August-2017, e-ISSN: 2348 -4470, print-ISSN: 2348-6406.M. P. Brown and K. Austin, *The New Physique*(Publisher Name, Publisher City, 2005), pp. 25–30.

IOP Conf. Series: Materials Science and Engineering 1170 (2021) 012014 doi:10.1088/1757-899X/1170/1/012014

Design and Development of Modular Arm Cast Using Additive Manufacturing

G Waghmare[1], S Bagde[2] and Sabir Khan[3]

[1] Mech. Engg. Dept.,YCCE, Nagpur, India
[2] Mech. Engg. Dept.,YCCE, Nagpur, India
[3] DPM, CPC Business Marketing Agency, Pune, India
E-mail: gwaghmare@gmail.com

Abstract : The plaster cast and fiberglass cast are the popular elbow dislocation treatment method for immobilization. These are relatively heavy, non-removable, and poorly ventilated, hence might create skin diseases, irritation under the cast and joint injury,if not cast properly. It takes long preparation time and cannot be adjusted after casting, which may lead to improper healing in case if the bone gets little shift insidethe cast. In this work, the orthopedics arm cast prototype is designed and fabricated by using additive manufacturing with ABS material. The core concept of the model is the elbow support and modular rings with a specific width added to achieve a specific support length. After properly positioning the patient's arm the pressure on the arm can be adjustsed by the small clamps with velcro straps. This has advantages over conventional cast that it provide cast adjustments, proper ventilation, washable, and reusable.

Keywords: Arm cast, modular rings, elbow immobilization, additive manufacturing

1. Introduction

An orthopedic plaster cast is a wrapped casing, made of bandage and glass wool, enveloping a fractured portion of the body to support broken bones for healing. The cast applied for the arm, wrist, and/or hand is called Upper extremity casts. The cast which envelopes the arm from the hand to about 2 inches beneath the armpit, retaining the fingers and thumbs exposed is called a long-arm cast. Similarly, the cast applied on shorter lengths on both sides of the elbow joint is termed as short arm cast.

Generally, the dislocation of the elbow occurs whenthe elbow joint is drawn apart. It can be partial or complete. Here we consider partial dislocation which can be referred to as a subluxation. The most common injury cause in elbow dislocation is impacted on the arm due to sudden fall. The impact force getstransferred to the elbow through the forearm. So dislocation of the elbow occurs.

Generally, Elbow recaptures the original position byits nature, however if not, the orthopaedic [1] surgeon aligns it by applying force. The orthopedic surgeon applies cast plaster on this joint and support. Around four weeks are required for a complete cure for these kinds of injuries. The typical plaster casting process of elbow support is given below:

The components required for plaster forming include loosely knitted stretch fabric, cotton wadding, plaster or fiberglass [2] roll or sheets, stretchy bandages with clips or adhering tape, high-quality scissors, and water. To start building up an elbow orthosis is to apply a protective layer using a loosely knitted stretch fabric. Then, for more comfort, the cotton wadding material iswrapped over the loosely knitted stretch fabric. The plaster sheets are soaked in water and applied over the wadding. The plaster is adjusted to the appropriateanatomical [3] position of the joint. In the last layer,the loosely knitted stretch fabric is wrapped with cotton wadding again and a stretchy bandage is used tosecure the plaster. This procedure takes about 45 to 50 min. This is mostly done in the clinic and hospitals. The patient has to wait until the preparation of the cast. Moreover, it cannot be adjusted or modified after curing, which leads to skin infection, irritation, and sometimes swelling as the patient cannot wash or clean skin inside the cast. In the conventional method, the cast must be cut off and destroyed after only one use and it is being done incorrectly, it may injure the patient.

2. Problem Identification

Presently, there are arm cast and splints that are also made from additive manufacturing, but this technique is lengthy. The time required for CT scan

or MRI of the patient's elbow and hand, thenpreprocessing with computer software, and finally the additive manufacturing of patient-specific cast will addup to painful waiting time for the patient. Although a specific cast for each patient is the best solution, the cost and time required are the constraints, also theavailability of all these systems under one roof is a major constraint. Due to these issues, there is a need ofdeveloping a modular arm cast for most of the patients having partial elbow or subluxation. The new concept should have attributes such as short preparation time, does not require specialist or skilled physician, adjustable and can also provide flexibility in adjustment during the healing process if required. The patient can be able to wash and clean his hand to keep himself away from irritation and infections, also provide proper ventilation. It will be designed in sucha way that it can be reused, which reduces the cost for the patients. [4]

The rapid prototyping method is used for the fabrication of a prototype. The CAD model file is converted to the .stl file. Then this is submitted for slicing. In this process, the drawing made in the CAD software is approximated by triangles and slice containing the information of each layer that is goingto be printed. [5]

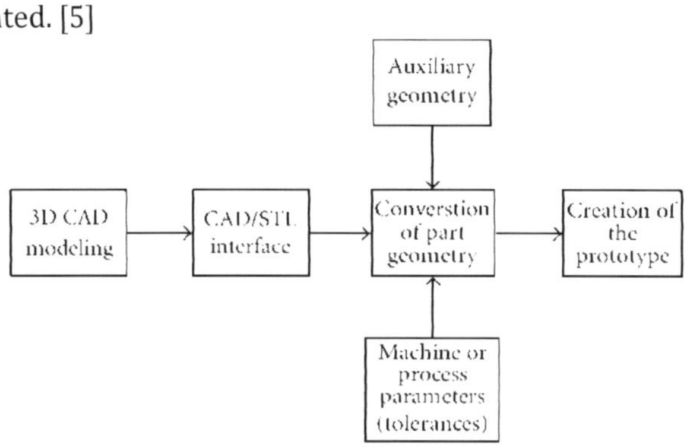

Figure 1. Additive manufacturing Procedure

3. Design Methodology :

Initially, the types of dislocation of the elbow are investigated, which are:

a) Dislocation of Humero-ulnar joint.

b) Dislocation of Humero-Radius Joint.

c) Dislocation of Radio-ulna Joint.

During dislocation of the elbow, the ligaments may gettorn. The ligaments act as cushioning for joints and also provide stability to joints. Proper healing of the joint will ensure natural alignment and stability. [6]

Table 1 shows the weight of the person and associated arm sizes at which the elbow joint can be fixed. The circumferential arm sizes at below and above elbow joint of samples of average Indian males/females. The size of the cast may vary at the particular weight ranges. Consequently, the size of the ring and elbows are divided into different sizes.

Table.1 Different size of the Arm according to theweight of the person

Weight (KG)	Arm sizes (CM)
40-55	21-25
50-65	24-32
60-80	25-34
70-90	27-36

The main concept is of changing the existing alternative of plaster or glass fiber cast to adjustable elbow arm cast. The model consists of elbow and Rings made of ABS plastic. Inside the ring, an arm holding clamps are provided better grip andpositioning. Below the ring's outer surface, the Velcro strips are given to adjust it properly according to the patient's elbow. More than one ring can be added as per the patient's elbow injury. [7]

At first, a ring along with clamps is prepared in 3D modeling SOLIDWORKS software. The arm size range of 24-32cm is taken, for the average weightrange of 50-65 kg. The ring circumference and elbow angle can be modified as the allowable angles of stabilization for the elbow dislocation are 0°-30°-90°- 110°-130°. After making a solid model of all parts and finishing it, thedata are translated into polygon STL format.

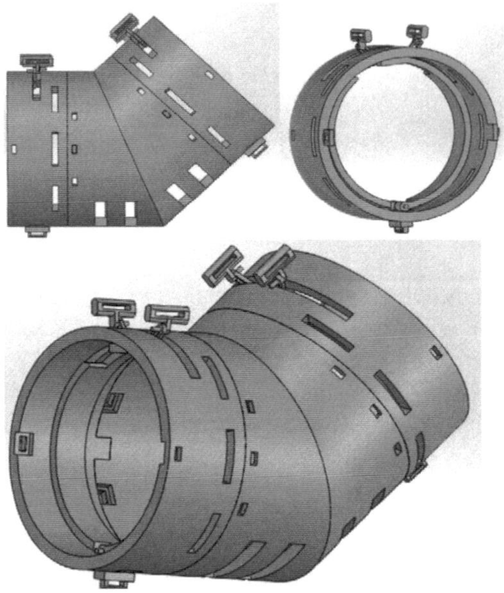

Figure 2 CAD model of Arm Cast

The fused Deposition Modeling method is used for prototyping after the STL file is made. We only used Acrylonitrile Butadiene Styrene (ABS) material for making the prototype of parts. Other medically accepted and skin-friendly material can also be used for manufacturing. [8] The material properties for the most suitable material for making cast are given in Table 2.

Table.2 Material Properties for FDM for the cast

Material	Desnsity gm/m3	Tensile Strength (MPa)	Flexural Strength (MPa)	Impact Strength J/m
1. (ABS) Acrylonitrile Butadiene Styrene	1.05	22	41	107
2. (PC) Polycarbonate	1.2	68	104	53
3. ABSi (Acrylonitrile Butadiene Styrene – Biocompatible)	1.08	37	62	96
4. PC-ABS (Blend of polycarbonate and ABS)	1.2	41	68	196
5.Nylon 12	1.01	32 MPa	67	135

The prototyping [9] method used here is Fused deposition modeling (FDM). In FDM, a thin plastic wire is fed to the printer head which melts it and extrudes to deposit in approximately 0.25 mm thickness. Medical-grade materials like ABS- Acrylonitrile butadiene styrene, [10] polycarbonate, and blends are used. This is the most economical method as no post-processing or curing of plastic is required. While SLS and SLM are also most preferred according to the accuracy and finishing of the components, but the cost of machine and prototypingis expensive as compare to FDM.

3.1 Wall Thickness of Section:

The outer layer is subjected to impact force exerted by external bodies and internal distortion force by arm movement. Hence it has to be made to withstand these forces. The external force of 200N is considered for analysis, assuming the speed and weight of the hand, during impact.

MSC *Nastran* software's structural analysis module isused to determine the wall thickness. The ABS material is a blend of plastomer and elastomer, exhibits flexibility, and hence can be used for intricate shapes.

The wall thickness of the ring plays a very important role in this structure. It should be as light in weight, strong and stiff, and provides comfort to the patient.

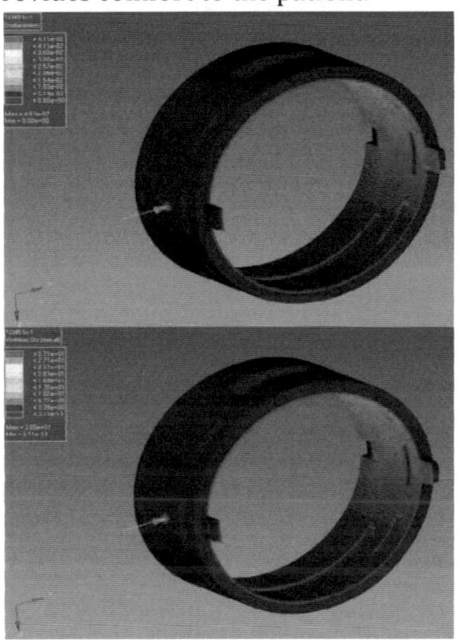

Figure 3 Structural analysis of the arm ring.

The main factor for influencing weight is wall thickness. From the analysis, the maximum displacement is found to be 0.04 at 200N force with a maximum stress of 30MPa. Hence the Finite element analysis determined the allowable thickness to withstand in forces and impact.

4. RESULTS AND CONCLUSION

The proposed concept for different sizes of arms cast can be designed and manufactured. In this work, we built the rings, elbow, clamps, and sliders using the FDM 3D printing technique. The manufactured parts are shown in figure 4. After the trials, significant results are observed as it fits well, provides good ventilation, support, and light in weight. The total weight of the proposed arm cast is 310 gram which is 1/3rd of the weight of a traditional plaster cast.

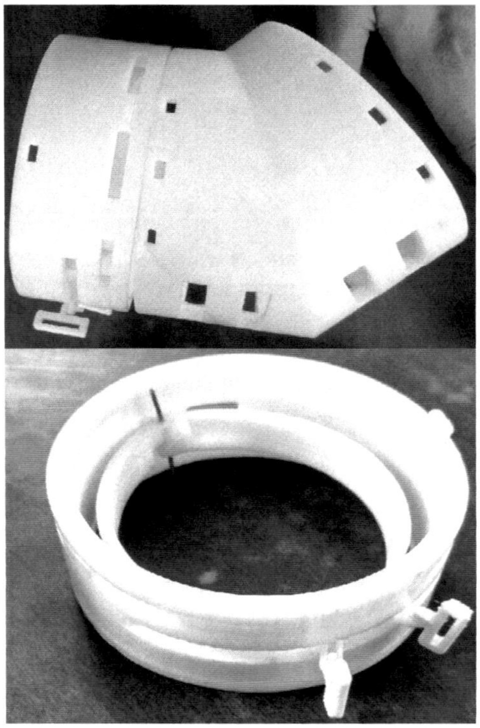

Figure.4 Additive manufactured RP of Armcast Elbow and attached Ring (top), Arm holding Ring with Clamps (bottom).

The modular arm cast prototype made using the FDM type of additive manufacturing is acceptable in thetreatment of dislocation of the elbow and can be a better substitute for conventional plaster. [11] This modular arm cast can be reused after sterilization.

This new cast consists of a modular ring that can be attached with the elbow cast on either side, to adjust the length requirement. The inside clamps will surround the arm and grip it by fastening the Velcro belt. This provides good support, ventilation [12], flexibility in assembly, ease in assembly. Thus this arm cast can be a better alternative to the traditional casting system. The modular cast is found to be lighter, well ventilated, durable, with adjustable lengths, and avoids fungal infection to the skin. This gives a better experience to the elbow dislocation patients, as compared with traditional plaster.

One of the main advantages of this proposed designed cast over a current patient-specific cast is that it provides reusability. This new design will give quick treatment to the patient having elbow dislocation for good flexibility to adjust and wash, even in a worn position.

5. FUTURE SCOPE

Little modification at the snap joint was required as theprojected part for the snap joint was too hard. Being thermoplastic material the shape can be adjusted with the use of glue or a hot gun. The POP plaster or glass wool plaster does not exhibit this modification. However, there is always a scope for further modifications for developing it as more robust andflexible. [13]

The following are the few suggestions for future studies:

1. A thorough study of the arm size and shapes of the human is required and accordingly, groups can be made and arm cast for respective groups can be fabricated.

2. The possibility of use for surgical injuries to be explored.

Finally, this work focused only on the size range of 24-32 circumference of the arm as further ranges will have to be tested. The inputs from various medical professionals and patients are required in the future
[14] for developing a more compatible and flexible design.

6. REFERENCES

[1] Auricchio, Ferdinando; Marconi, Stefania (2016). 3Dprinting: clinical applications in orthopaedics and traumatology. EFORT Open Reviews, 1(5), 121–127.

[2] Angus P Fitzpatrick, Mazher Iqbal Mohanned, Paul K Collins, and Ian Gibson, (2017), Design of a Patient- Specific, 3D printed Arm Cast, The International Conference on Design and Technology (2017), Volume 2017,135-142.

[3] Paterson, Abby Megan; Bibb, Richard; Campbell, R. Ian;Bingham, Guy (2015). Comparing additive manufacturing technologies for customised wrist splints. Rapid Prototyping Journal, 21(3), 230–243.

[4] Fitch, Michael T.; Nicks, Bret A.; Pariyadath, Manoj; McGinnis, Henderson D.; Manthey, David E. (2008). Basic Splinting Techniques. New England Journal of Medicine, 359(26), e32–.

[5] Wong, Kaufui V.; Hernandez, Aldo (2012). A Review of Additive Manufacturing. ISRN Mechanical Engineering, 2012(), 1–10.

[6] Matthew Halanski, Kenneth J. Noonan,(2008) Cast and Splint Immobilization: Complications, Journal of the American Academy of Orthopaedic Surgeons Volume 16
- Issue 1 - p 30-40

[7] Englert, Carsten; Zellner, Johannes; Koller, Michael; Nerlich, Michael; Lenich, Andreas (2013). Elbow Dislocations: A Review Ranging from Soft Tissue Injuries to Complex Elbow Fracture Dislocations. Advances in Orthopedics, 2013(), 1–11.

[8] Kim, Huhn; Jeong, Seongwon (2015). Case study: Hybrid model for the customized wrist orthosis using 3D printing. Journal of Mechanical Science and Technology, 29(12), 5151–5156

[9] Potamianos, P.; Amis, A. A.; Forester, A. J.; McGurk, M.; Bircher, M. (1998). Rapid prototyping for orthopaedic surgery. Proceedings of the Institution of Mechanical Engineers, Part H: Journal of Engineering in Medicine, 212(5), 383–393.

[10] Sutradhar, Alok; Park, Jaejong; Carrau, Diana; Miller, Michael J. (2014). Experimental validation of 3D printed patient-specific implants using digital image correlation and finite element analysis. Computers in Biology and Medicine, 52(), 8–17.

[11] Eltorai, A. E. M., Nguyen, E., & Daniels, A. H. (2015). Three-Dimensional Printing in Orthopedic Surgery. Orthopedics, 38(11), 684–687

[12] KELLY, S., PATERSON, A. and BIBB, R., 2015. A review of wrist splint designs for additive manufacture. IN: Proceedings of 2015 14th Rapid Design, Prototyping and Manufacture conference (RDPM 14), Loughborough, Great Britain, 15-16 December 2015.

[13] Mohd. Javaid, Abid Haleem,(2018) Additive manufacturing applications in medical cases: A literature-based review, Alexandria Journal of Medicine,
Volume 54, Issue 4, Pages 411-422

[14] Mulford, Jonathan S.; Babazadeh, Sina; Mackay, Neil (2016). Three-dimensional printing in orthopaedic surgery: review of current and future applications. ANZ Journal of Surgery, Sep;86(9):648-53

AUTHOR INDEX